FIRE!

ON THE SCENE IN NASHVILLE

Gordon and Martha Hall

The Overmountain Press
JOHNSON CITY, TENNESSEE

Front cover photograph by **Rick Musacchio / *The Tennessean***. Metro Nashville firefighters battle a blaze that heavily damaged an abandoned house in east Nashville.

ISBN 0-932807-62-3

Printed in the United States of America

1 2 3 4 5 6 7 8 9 0

FOREWORD

Few people realize the dangers of a fireman's life. Dangers of every description are met unflinchingly by the man of the hose—yet some people think he has "a cinch" when it comes to a job. His work never ceases. Twenty-four hours a day he is on duty, never knowing what moment he will be awakened from a sound sleep to rush off in the dead of night to fight the flames or on a rescue mission. Firemen are called on in every kind of emergency to perform unheard-of feats.

When a fireman retires, his boots are left standing by his bed, his trousers in position over them, so that when the alarm rings he can get dressed quickly. They pay little heed to personal safety. The hardships that go with fighting fires are a hundredfold. Soaking wet, freezing cold, choked with smoke, burning hot, bruised by falling timbers and begrimed, he must stick to his post until the last spark is dead. Acting on orders from his superiors he goes into a burning building, the floors falling beneath his feet, carrying the hose to a place where he can play on the flames to advantage.

On the days when there are few fires, the firemen work at the fire hall. Windows are washed, fire hall cleaned, apparatus cleaned and shined, records kept, study and exercise programs must be attended. The grounds are kept clean.

Today's firefighter must know about hydraulics if he is to make intelligent use of water streams in extinguishment of fire; about chemistry, if he is to fight fire with chemicals; about law enforcement, if he is to conduct fire prevention and fire protection programs. Along with these he must know all phases of fire fighting, evolutions, first aid, basic life support, emergency medical service and rescue work.

To relieve the tension of always being on the alert they do see humor in some situations and also play practical jokes on one another.

We appreciate the firemen of Nashville sharing their stories—some tragic, some funny and some heart warming. You will laugh and cry and love with them as we have over the years.

In spite of the danger and hardships, firemen are dedicated to the service of their fellowman. *No greater love has any man than to lay down his life for his friends.*

FIRE SPEAKS

I am a vassal of man and I do his will
  In many a marvelous way:
If he chains me sure I am fain to endure
  His mastery night and day;
But should I escape from my dungeon red,
  To follow where embers fly,
I chasten my Master with Hell's disaster,
  And flaunt my head to the sky.

O.H. Roesner-*N.F.P.A. Magazine* 1916

PREFACE

This book was written for several different reasons; the first being my husband, Gordon W. Hall, was a fireman with the Nashville Fire Department for thirty-four years. He retired January 1, 1990. As all firefighters and their families know, firefighting is a way of life that involves and affects the whole family's lifestyle. Gordon and I decided to record for our family and other firefighters' families a part of the Nashville Fire Department as we knew it.

We have received invaluable help and wonderful cooperation from Nashville firefighters, the Nashville Fire Department, the Nashville Archives Library, the Ben West Library, *The Tennessean, The Banner,* The Woodring Report, the Fire Buffs Club, Tennessee State Archives and the Metro Archives.

A special thanks to Evelyn Wright, Sue Gossage and JoAnna Long who helped with the editing. And special thanks also to Randal Hall and Vonnie Hall who helped with the computer work.

Our appreciation to our children, Randal, Steven, Vonnie, Janet, Bruce, Donnie and Kathy, and our grandchildren who lived a firefighter's family life and encouraged and loved us all the way.

Gordon and Martha Hall

FIREMAN'S PRAYER

When I am called to duty, God
Wherever flames may rage,
Give me the strength to save some life
Whatever be its age.
Help me embrace a little child
Before it is too late,
Or save an older person from
The horror of that fate.
Enable me to be alert and hear the weakest shout
And quickly and efficiently
To put the fire out.
I want to fill my calling and
To give the best in me,
To guard my every neighbor and
Protect his property.
And if according to my fate
I am to lose my life,
Please bless with your protecting hand
My children and my wife.

Anonymous

TABLE OF CONTENTS

Foreword / iii
Fire Speaks / iv
Preface / v
Fireman's Prayer / vi

1 History / 1
The East Nashville Fire / 7
Arrival of the Fire Trucks / 18
Our Fire Department / 19
The Maxwell House / 20
Most Beautiful Fire / 27
The Fire at the Fairgrounds / 28
State Fairgrounds Fire / 33
Radio Communications Replaced / 34
A History of the Nashville Fire Buffs / 36
This Is the Fire Department / 38

2 Stories by Firemen / 41
Stuck! / 41
Oh! Not Now! / 41
A Tale Twice Told / 42
Be Patient Your Time Will Come / 44
A Real Cool Fireman / 44
She Got All Mad / 45
My Dog Story / 45
Too Late / 46
A Gas Mileage Problem / 47
Round and Round She Goes / 48
I'll Get You / 49
My First Fire Call / 49
I Couldn't Have Done It / 50
A Memoir / 51
Too Old and Too Slow / 51
Bugged / 51
They Taught Me to Be on Time / 52
Shake It Hard / 53
You Lousy Firemen / 53
A Green Bucket and It's Yours / 54
A Slick Trick / 55
She Did Not Forget / 56

The Mystery of the False Alarms / 57
An Accident That Should Not Have Happened / 57
Better to Leave / 58
Firemen Can Do Anything / 58
He Found It / 59
Frozen In / 59
Call the Chief Off / 60
An Unsuspecting Good Old Man / 60
What a Night / 61
Some Friend / 61
Help Me Somebody / 62
No Water—No Line—Nothing / 62
Snakes Alive / 63
I Just Can't Do It / 63
Jingle! Jingle! / 64
Go Ahead / 65
We Gotta Eat / 66
An Unwelcome Guest / 66
My First Three Days / 67
Deadly Gas / 69
A Retired Fireman Reflects Thankfully / 69
No Ponds For Us / 70
Your Decision / 71
You Can Get Paid Twice / 72
Snow And More Snow / 72
Where Are the Nuts? / 73
Sad Memories / 73
Scared of a Kitten / 74
Fire Prevention / 75
Memories / 76
Harvey's Had It / 77
You Can't Work Here / 78
Watch My Coffee / 79

3 Animals / 81
Sparky / 81
Spot / 82
Echo / 83
Fire Hall Dan / 84
Pooch / 85
Roy / 86

4 The Last Alarm / 87

C.C. Gowdy / 87
Stanley Allen / 87
Harvey Ewing / 87
R.E. Page / 89
E.M. Carell / 90
R.C. Burk / 92
George Rogers / 94
William R. Green / 96
J.C. (Bud) Ambrose / 96
Wallace Pinkerton / 99
A.A. Rozetta / 99
Thurmond Oscar Dillard / 101
John S. (Penny) Pennington / 102
Johnson Van Tien / 103
W.H. Norris / 104
Truman Hill / 104
F.R. (Buddy) Reece / 104
Joe Keef / 106
Thomas Madden / 106
Charles Oberchain / 106
Ben Christian / 109
Ed Dale / 110
Louis R. Ryman / 110
Max Solinski / 110
S.B. (Shad) Sullins / 110
Horace Overholt / 111
Raymond Goodrich / 112
Joe Knight / 112
Bernard Mattingly / 113
Wylie M. Carell / 113
Orville Hooberry / 114
J.H. Cooper / 114
Raymond Parrish / 114
Charles Coe / 114
George H. (Hub) Stephens / 115
Morris Shutt / 116
Vernon Keel / 116
Thurmond Gentry / 116
James Edward Montgomery / 116
John Hall / 117
Charles Lovell / 117
Nameless Heroes / 117

1 History

EARLY PHOTOGRAPHS OF THE NASHVILLE FIRE DEPARTMENT — LATE 1800s AND EARLY 1900s

(Courtesy of Tennessee State Archives)

Early horse-drawn engine ca. 1895. Firemen unidentified.

Truck Company 2.

First car used by Nashville Fire Department in 1909, made in Nashville by the Marathon Motor Works.

Early steam engine ca. 1885, currently on display at Opryland

Horses (Ben, Pete and Mack) of Hook and Ladder Company 1, 1890. Driver not identified.

Chief Rozetta and unidentified driver in first fire car.

Photo of Engine Company 14, 1916. This fire hall still in operation today.

Engine Company 6 ca. 1895.

Fire truck #3 ca. 1927

Hose reel in front of Fire Department Headquarters ca. 1895.

Engine built by Nashville Fire Department repair shop ca. 1895.

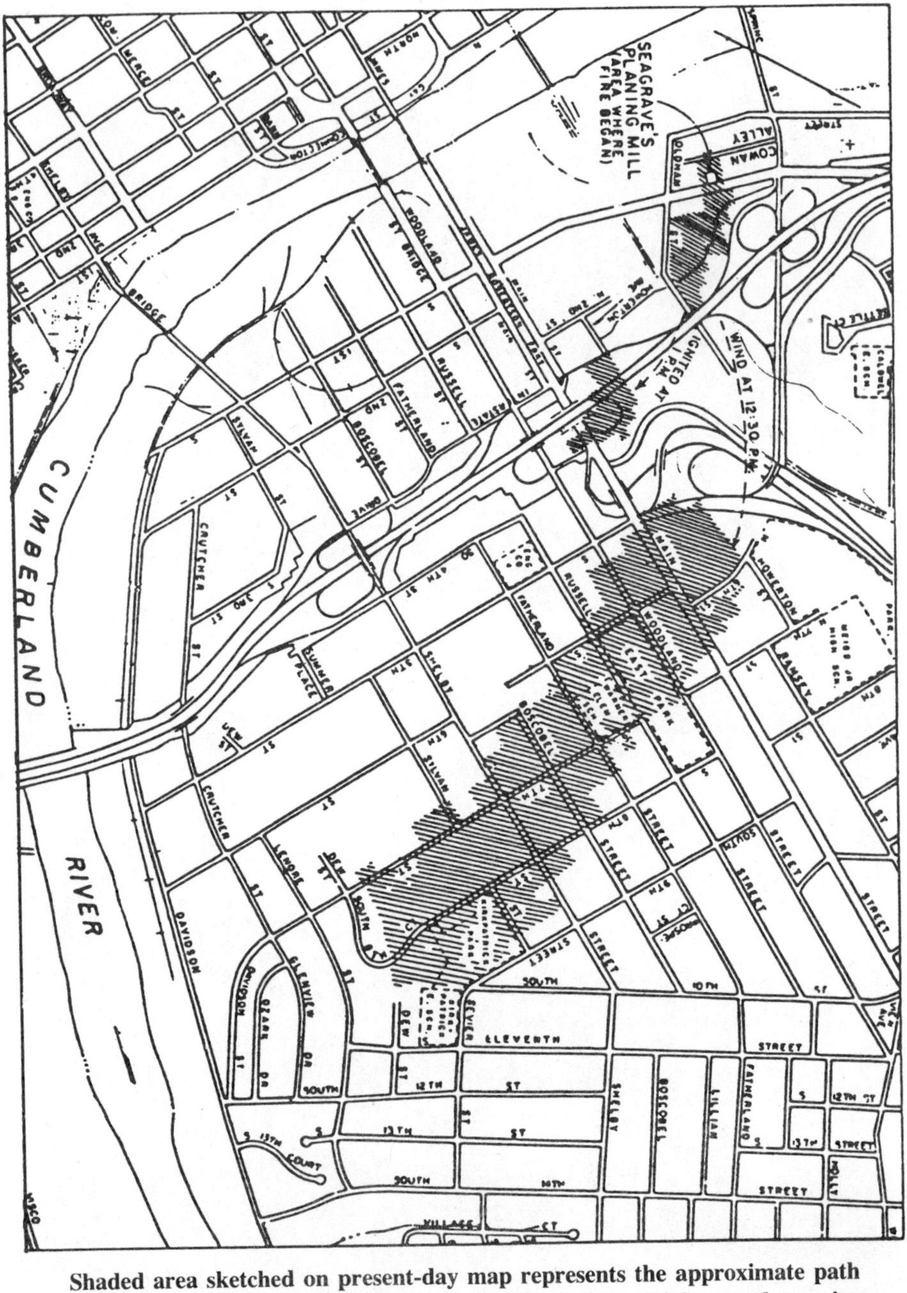

Shaded area sketched on present-day map represents the approximate path and extent of destruction by the fire, which raged for 4½ hours, destroying 32 city blocks with 648 buildings. (Courtesy of Historic Nashville)

THE EAST NASHVILLE FIRE

"God Almighty in his wisdom often makes us feel our littleness. He shows us up in our weakness. Wednesday I felt the size of an ant," Chief A. A. Rozetta said as he recalled the horrifying events of Wednesday, March 22, 1916.

He was as well prepared as premonition would allow him to be, for as the wind began to rise on Tuesday night he had grown increasingly nervous. As he paced the floor he warned his men, "I feel like we are in for a big one."

Premonition and no amount of preparedness humanly possible could deal with the facts he would encounter on that fateful day. A wind that reached the velocity of from thirty to sixty-eight miles per hour, the dry conditions of grass and timber, the light construction of the houses, the wood shingle roofs and open foundations of the homes that were in the path of the fire all added up to a holocaust unleashed on the city that proved uncontrollable.

When he awoke Wednesday morning the wind was still howling at cyclonic proportions. Still feeling the nagging foreboding in his heart, he called all captains throughout the city instructing them to keep every man on duty and to have all equipment in readiness in case a fire broke out. So strong was his conviction that he refused one captain permission to take an engine into the shop for minor repairs.

Earlier that year he had warned the mayor and other city officials that East Nashville had grown so quickly that the citizens were in a very precarious position. Only a few fire hydrants existed. Some of these had low water pressure. Firemen were often forced to rely on the cisterns of private citizens for water supply to fight fires. The light construction, wood shingle roofs and the open foundations made these homes and businesses like tinder boxes.

On that Wednesday morning Chief Rozetta was at the Dixie Chemical Company with Engine Company 4, fighting a dump fire only 600 yards from the origin of the blaze that was to become known as the East Nashville Fire, the most disastrous fire in the history of the city.

At 11:47 a.m. the firemen at the dump were horrified to see a huge, sinister, black cloud rolling from the roof of the Seagraves and Company Planing Mill at 214 North First Street. Realizing immediately their futile plight in the face of the fifty-one-miles-per-hour wind blowing at that time, Fred Cassety ran to pull the box alarm at the corner of North First and Oldham Streets.

Engine Companies 1, 2, 3, 4, 9 and 10 and Truck Companies 1 and 2 responded and were immediately put into service. The fury of the wind whipped the sparks southward. Six houses exploded into flames before the fire equipment could be moved the 600 yards. Engine Company 4 frantically hooked up to the fire hydrant and found 80 to 90 pounds of pressure. On an ordinary day it would have been plenty, but this was no ordinary day.

In a matter of minutes 33 buildings in that immediate vicinity were ablaze.

Looking southeastward, veteran Chief Rozetta spotted the very sight that his heart had dreaded. His head flew back and every muscle in his body grew tense. With a leap he was in the driver's seat of his red car speeding up Oldham Street. Two blocks away on Howerton Street he saw a roof on fire. Three blocks away to the east on Donelson Avenue there were several small cottages being quickly consumed by the flames. The wind was blowing with such force that it had carried burning embers across 1800 feet of cleared area along the railroad tracks to the wood shingle roofs of the thickly populated area.

Fred Cassety tried in vain to turn in a general alarm from Box 76 but it was dead. Electrical wires had burned in two and were falling all around him. He ran about one-third of a mile to Box 71 at Woodland and Bridge Streets with the urgent message to send all help available. This general alarm was received at 12 noon. It was the first time Nashville had experienced a general alarm.

In the few minutes it took Chief Rozetta to circle back to Oldham Street, his campaign had been outlined. He gave the orders to his captains, barely slowing down his automobile as he passed.

The front of the fire by this time was four blocks wide, from 3rd to 7th Streets. Only with heroic work on the part of his men, and immediate action, could the line be narrowed to two blocks—or perhaps to only one.

Within fifteen minutes all companies were in their positions assigned by the Chief, and the fight was on to hold the flames within as narrow a pathway as possible. And hold them they did for 35 blocks—for more than 2½ miles the width of the fire did not exceed the two blocks.

History has proven that the decision to try to contain the fire in this narrow pathway was the wisest decision that could have been made.

The boundaries of the fire were not fixed by streets but diagonally across squares, making the work of the department more difficult in the effort to hold it within a prescribed pathway. The men were also hampered by falling telephone cables, 2300 primary lighting wires and 500-volt trolley feeders. In the path of the flames, nearly every pole burned to the ground. A total of 1500 feet of fire hose was destroyed. The firemen were hampered to some extent as a result of having to replace burned sections of hose while fighting the flames. Picking up momentum and roaring like a prairie fire, the inferno swept quickly through the flimsily constructed homes, leaving only charred ruins and blackened chimneys standing where only minutes before people had been going about their usual noontime chores.

The brave men of Engine Company 5 from McLemore Street made a valiant stand between 4th and 5th on Main Street. Inch by inch they fought the searing wall of fire. Their determination was heightened by the fact that just ahead of the fire was the Little Sisters of the Poor home for the indigent and infirm. There, too, stood St. Columba School, filled with children, and St. Columba Catholic Church. When intense heat, dense black smoke and exhaustion forced them to give ground, the firemen found they did not have time to hook the horses to the finest horse-drawn engine in the department. So fierce was the blaze that it burned the wheels off the engine.

After the firemen were forced to give up their stand, a most pitiable situation developed as the roaring inferno approached the beautiful three-story brick building of the Little Sisters of the Poor at 521 Main Street. Housed there were 48 elderly and infirm men and women. Hoping against hope that their building would be spared, the sisters had delayed the evacuation as long as possible. But when all hope was gone the evacuation began. Police and volunteer citizens helped remove the patients. Most of them were hysterical with fright. Many begged to be left alone, believing that this was their only place of safety. One lady who had lived there for years pleaded to be left to burn with the building because it was the only real home she had ever known. Dazed and incoherent patients were found wandering around on the street. Most of them had to be forcibly put in private automobiles and taken to the basement of St. Mary's Church, where the National Guard provided cots for the night.

The flames leaped across Main Street and soon devoured St. Columba Catholic Church and school, leaving only a pile of rubble.

On the flyleaf of the baptismal register, Father Eugene Gazzo made this entry: "March 22, 1916. Church, school and residence with all contents reduced to ashes in the great conflagration. So passes away the glory of the world. The Lord has given and the Lord has taken away: blessed be the name of the Lord. The mercy of the Lord that we were not consumed."

Other beautiful and historic churches destroyed were the Woodland Street Christian Church, the Woodland Street Presbyterian Church, the Tulip Street Methodist Church and the Edgefield Baptist Church.

With every company of the fire department in the city, every fireman in service and most members of the police and detective departments on the scene, urgent pleas were wired to Evansville, Louisville and Chattanooga for help. It was expected that fire engines from all three cities would arrive as soon as the fastest trains could bring them.

Meanwhile, the wind picked up burning pieces of framework and hurled them as far as six blocks through the air to the roofs of other houses. The fire ate its way from building to building into cross streets and took in all houses in several blocks. The fire raced through the expensive homes on Woodland Street; there were fourteen burning at one time. It roared from Woodland to Fatherland, across Fatherland, on to Russell and across Russell to Boscobel, across Boscobel to Shelby and into the streets that crossed, then across Shelby. Chief Rozetta said: "There was one time when it appeared that we were going to lose in our fight to keep the fire in the pathway. I felt that it was going to get away from us, and that was when it reached Shelby Avenue. The water pressure in the pipes gave out by the time we reached the Shelby Avenue grade. We were forced to rely wholly on the engine pressure. Everything was concentrated, however, and with heroic work on the part of the men we kept it from spreading."

The fire was advancing at the rate of 42 feet per minute. It destroyed more than two buildings each minute. The avalanche of destruction tore through on to Sylvan, Sevier and Pitt Streets and finally ended at Lenore Street, near the Cumberland River, leaving sixty-four square blocks of devastation.

As the thousands of residents saw the fire approaching, they took every precaution available. Neighbors helped neighbors with bucket brigades and garden hoses; they even placed water-soaked quilts on roofs to try to save their property.

As the fire destroyed the expensive homes on Woodland Street, it became apparent that the beautiful Warner School building at 7th and Russell Streets was doomed. Warner, a 15-year-old three-story building, was the second largest school in the city, with 21 teachers and 900 students. It was a fearful decision that was forced upon Principal Finis K. Henderson to dismiss these children to face the horrible confusion all around them. With the flames only a short distance away, the children left the building in an orderly fashion and none were injured. Within a few minutes the school was destroyed. Only the walls of the 100-foot clock tower were left standing.

The fear and hysteria mounted as families stood and watched everything they had go up in smoke. In many homes not a thing was saved. One woman who begged and pleaded to go back into her home was forcibly placed in an auto. Still screaming, she was carried away from the scene.

There was constant prayer in the hearts of firemen and residents that the wind would shift or die away altogether, but the elements seemed to take a delight in stirring up the fury of the conflagration.

Automobiles, wagons and buggies that were not being used as ambulances to transport the injured to hospitals were used to take household belongings out of the fire zone.

With the roar of wind and fire, the hysterical cries of citizens, and the wail of ambulances and fire engines, the sound was maddening.

Thousands of people were piling their household goods in the streets. Preparations were started early in the afternoon to protect the belongings from thieves. Chief of Police Alex Barthell sent every available patrolman and plain-clothes officer to the scene. When the enormity of the situation was realized, Governor Rye was notified. He immediately issued a call for the National Guard. Within an hour, fifty members of the 2nd Battalion were leaving the armory for East Nashville. By seven o'clock more than 100 had reported to Engine Company 4 on Woodland Street. In addition 100 volunteers were given commissions to do patrol duty.

An announcement was made in *The Banner* newspaper that volunteers were needed to assist police in overnight guard duty. Applicants who were deemed worthy were given a cloth badge bearing the word "Police" with a pin for fastening it onto their coats. No effort was made to swear in the volunteers. Wearers of the badges were permitted to pass the military lines unquestioned.

Within four hours and 45 minutes an area of approximately 64 acres had been devastated. The fire destroyed 700 buildings with an estimated value of close to two million dollars, by that day's standard. Three thousand people had been left homeless. Homes, churches, schools and private businesses—all were leveled.

Records show that loss of life was limited to one or possibly two persons—through the heroic efforts of firemen, police and volunteer citizens' warnings to leave the fire zone.

The motherly instinct of animals was graphically portrayed in a story told by Patrolman J. K. Bradford. He was assigned to keep in advance of the rapidly spreading fire to warn the citizens of the districts known to be doomed. "On Patterson Street there was a large newly plowed garden," said Mr. Bradford. "Though the buildings on every side were a mass of flames, an old mother hen carried her baby chicks to the very center of this plowed ground and hovered closely over them. At times it appeared that the blaze would literally sweep her away, but she held her ground and the chicks remained sedately under her. When the worst had passed, the old hen was seen again. Although blackened by falling ash and dirt, her head was high. She had already begun the task of finding supper for her brood."

Too much cannot be said in praise of the firemen who stayed at their posts of duty until overcome with exhaustion. Under the direction of Chief Rozetta and Assistant Chief Weaver, they worked feverishly until many were overcome by heat and heavy smoke. They pursued their duty against what seemed to be an uncontrollable situation. Many members of the department suffered slight burns; some were taken from their posts completely overcome.

As nightfall came on, the entire section east of the river was plunged into total darkness, without telephone or trolley service.

Massive efforts to relieve the suffering of the victims of the fire came pouring in from all parts of the city. Fellow Nashvillians hurried to the stricken to provide food and clothing. Many citizens opened their doors to the 3,000 who needed shelter that night.

The following day, strenuous efforts began to restore all utilities and trolley service. Insurance claims were quickly rendered in the next few days.

The day after the fire, an investigation by C. W. Shuyler, Commissioner of the Department of Fire Prevention, showed that the fire had

PHOTOGRAPHS OF THE EAST NASHVILLE FIRE
(Courtesy of Tennessee State Archives)

started when a boy snatched a burning yarn ball from a grate and threw it out the door. Dried grass in the yard ignited. The cabin blazed up, and within a few minutes the flame had spread to a pile of shavings at the rear of the planing mill of Seagraves and Company. The sparks and embers from the mill were carried for blocks by the high wind. Then flames broke out several squares away. The fire department had been helpless in its efforts to stop the flames; and within 30 minutes after the little boy made such a desperate effort to save his toy, the hope that East Nashville could be saved was given up.

As daylight broke on March 23, the city's four bridges carried a constant stream of pedestrians and vehicles to East Nashville. Refugees, property owners, friends of the stricken, and the curious wound through the streets of the fire district, each with a different interest. The veritable forest of standing chimneys represented lost homes to some, the sorrow of lost friends and neighbors to others. And to many, they were reminders of the helplessness of man.

The largest crowd, estimated at 10,000, came to witness the dynamiting of the Warner School tower. The tower stood more than 100 feet high and appeared to be ready to fall. Early Thursday morning, Commissioner George Tompkins of the fire department had the walls tested. When they were declared to be dangerous, he ordered them demolished. Fifteen holes had been drilled at the base of the tower, and 600 pounds of dynamite were tamped into place.

The alarm was given at 3:30. The fuse was lighted. A minute later the spectators saw a puff of white smoke issue from the base of the structure. The puff was accompanied by the roar of the explosion. In a flash, the walls of the tower collapsed. It appeared as if the earth had opened up and swallowed the once magnificent entrance to Warner School.

The force of the explosion broke every brick in the structure, and the debris at the base of the tower became a mass of powdered sand and lime. The force of the explosion hurled broken bits of brick and dust for more than a square block.

While the fire-swept district was still smoldering, talk was heard on every hand among people who had lost everything they owned. They spoke of remaining loyal to the section in which most of them had lived for years. The slogan, "Stay in East Nashville," was growing increasingly popular. Still dazed from the catastrophe, many were already making plans for restoration. Old people who had come to East Nashville long before, and who had been active in its growth, began to plan to rebuild. To these grief-stricken people, home would not be home were it not in East Nashville.

ARRIVAL OF FIRE TRUCKS

On November 17, 1912, three new auto trucks arrived at the Nashville Fire Department. There was much concern and discussion as to the fate of the faithful old fire horses which these trucks would replace. It was Chief A. A. Rozetta's unhappy task to choose which of these faithful animals would be laid off.

Chief Rozetta did not take his task lightly. "We love our horses," he said simply. "They are almost human to us, especially to the drivers. When they go out of service, it's like taking their life away from them. They frequently die soon afterwards, even though given the best attention. It's just like taking an occupation away from a man. Nothing left to live for, so they die. Take the case of Warren. He was one of our best horses, but he got old and his legs got weak. When their legs get weak that's the end. You see you can't trust a great fire truck to a weak-legged horse in time of an emergency. They might fall down and a great loss of property or life would be the result of the failure of the engine to get to the fire on time." The Chief began to check his list and went on, "It's certain that Jumbo's got to go. He's over in Engine Company 2. Then there's Frank. We got him in '94. Frank is a dark bay. He is growing feeble and his legs are apt to give down. I'll relieve Frank from regular service this afternoon. And then there's Atkins. I guess Atkins had better go, too. Atkins is like Jumbo and Frank--a dark bay, named after an old blacksmith that used to shoe him. Hate to see Atkins go. He's been a mighty faithful horse. Then there's Canfield and Emmett over at the Chemical Engine House. Both of these horses' feet are in bad shape. I'm kind of hoping if we put them out on pasture for a while they might heal. I hate to give them up. Taking it all and all we haven't got a more faithful team in the service than they are. Buck, Rowdy and Ponce are growing pretty old and they will have to be laid off soon," he added. "But they are all good horses and an effort will be made to find work for them in a capacity they are able to fill."

Two horses named Hilary and White-Eye had bad reputations. Hilary especially would kick and fight every man who went near him, with the exception of his driver. "But when Hilary gets hitched up and on a run to a fire, he is one of the best pullers. They are hanging on to that Hilary."

The horses were eventually replaced by shiny big fire trucks. But there was not a man in the Nashville Fire Department who wasn't sorry to see the horses go.

OUR FIRE DEPARTMENT

To the Editor of *The Banner:*

Since the discovery of our city's financial condition there is considerable talk as to what should be done in the way of retrenchment. Quite a number of our good people are coming forth with the argument that our city employees should have their wages reduced. Some of them have singled out the fire department as one place to begin, claiming that the firemen are overpaid for the duties they perform. I, as a citizen and taxpayer, happen to be somewhat familiar with our fire department, and want to ask our good newspaper, *The Banner,* the privilege of expressing myself through its columns.

To begin with, the men in our fire department receive a salary of $90 per month for their services. They give twenty-four hours for a day's work. They are kept on duty eight days on a stretch before they are allowed to absent themselves from their posts of duty, being allowed only one hour to partake of their meals.

Now, Mr. Public, let's be fair. There is hardly any of us ordinary people or wage earners as I shall put it, but gets away from his labors at the close of day and most all of us get the seventh day of the week for rest and worship. At any rate, we are relieved of our responsibilities, that is, the majority of us are. Again, I say, let's be fair and consider what a sacrifice these men are making. I feel safe in saying, and I am in a position to know, that quite a good many of these men are religiously inclined, and would be glad to do something for their souls on the Sabbath day, but owing to the fact that they are so closely confined to their duties they must deny themselves of this privilege. Another great sacrifice they make, I will mention, is this: They are kept away from their families practically all of the time. Just stop and think what a sacrifice this must be. Most all of us have the privilege of spending half of our time at home with our families and loved ones.

If there is a deserving set of men on our city payroll, who earn every penny they get and more, it is our city firemen. These men are like most all of us, they have to live and have many obligations before them. Quite a number of them are buying their homes and pay taxes. Why should they be made the victims of our city officials' squandering acts?

We should be proud of our fire department, as it is second to that of no city of our size, as has been repeatedly demonstrated.

You commissioners who are going to handle the situation, I appeal to you in the name of justice to proceed with caution and, for heaven's sake, to not destroy everything it has taken us all these years to perfect. Our fire department is giving us a hundred cents on every dollar.

D. M. Gordon *(Tennessee Nashville Banner,* August 23, 1915)

THE MAXWELL HOUSE

"It's a total loss! A total loss!" These words of exhausted, emotionally drained Assistant Fire Chief Russell Campbell echoed throughout Nashville, across the nation and even internationally, as word spread that the historic Maxwell House Hotel in Nashville, Tennessee, now lay in a monstrous pile of smoldering ruins—victim of a tragic fire Christmas night in 1961.

Tragedy, war, conflict and fire had stalked the Maxwell House from its very beginning, but the hotel had gained worldwide fame for its beauty, genuine Southern hospitality and fabulous cuisine. Since President Theodore Roosevelt had asked for another cup of its coffee and made the off-the-cuff remark, "It's good to the last drop," its name had been carried far and wide by the popular brand Maxwell House coffee.

Once called "Overton's Folly," the hotel was born in tragic time. Col. John Overton bought the land, a grassy cow pasture on the edge of a growing frontier town, for $15.00. He thought he had bought a cow but, when instructing the auctioneer where to deliver the cow, he found that he had purchased a lot.

Plans were drawn up and ground was broken for the hotel on August 16, 1859. Col. Overton used slave labor to make the bricks at his home, "Traveler's Rest," located six miles from the hotel site. Slave labor was also used to lay the bricks of the solid wall structure.

The walls were up, the roof on, and the two top floors partially completed when the Civil War threatened to rip the nation apart. Col. Overton was elected commander of a regiment of militia and outfitted a company of volunteers. He went away to war leaving his building to the mercy of the warring factions, unfinished and unprotected.

Known then as Zollicoffer's Barracks, it was first used by the Confederacy to store cannons and to quarter soldiers. When Nashville fell captive to Federal troops the structure was commandeered to become a prison hospital. Housed there on the partially completed fourth and fifth floors were four hundred Confederate soldiers, most of them captured at the Battle of Chickamauga.

At eight o'clock on a hot, dry day, September 29, 1863, a most horrible catastrophe happened. When the signal was given for breakfast, more than two hundred hungry prisoners hurried to go down the only stairway. This stairway was temporary, intended for use only by carpenters in construction of the building. The soldiers were scattered along the temporary steps from the top floor to the second, when a Union soldier below called them to a halt.

Evidently the shift in rhythm of the marching men, plus the added weight of men who kept crowding onto the stairs brought on the accordion-like folding of the stairs.

This account of what happened appeared in *The Nashville Daily Union*: "The steps leading from the fourth to the fifth story gave way and fell with the men on them to the floor immediately underneath which also gave way as did also the floor next in succession and together with two flights of steps the whole falling in mass on the second floor!"

The screams of the dying men and the rush of the ambulance wagons alerted the citizens of Nashville. Vehicles from all over town rushed toward the prison. Weeping women with baskets of bandages pleaded to go inside to aid the horribly maimed and crushed prisoners. They expressed humiliation, bitter sorrow and helpless rage when they were held back by fixed bayonets of the Federal troops.

The Provost Marshal ordered the barracks cleared except for those who were clearing the wreckage or removing the dead and wounded.

The wounded men in excruciating pain were taken to military hospitals, some even out of state. According to one survivor of the disaster, the late John M. Dickey, 44th Tennessee Regiment of Kelso, Lincoln County, Tennessee: "There were 126 of us in the fall. Forty-five were killed outright or died in a short time."

The "Union Daily" said that six died immediately, ninety-six were wounded and, of these, fifteen were badly hurt and would probably die.

In those days Federal authorities controlled local papers. The disaster was given scant publicity, and statistics on the disaster vary.

After the war was over Col. Overton returned to Nashville. Although many claims for property damaged by the troops of the occupation were paid, Col. Overton was never paid for the abuse of his building.

After eleven long tedious years from its beginning, the hotel was completed at a cost of a quarter of a million dollars—a great sum in those days.

Whispers of "Overton's Folly" did continue to circulate. Many wondered, "Could this small but growing frontier town support this huge, luxurious hotel?" Few dreamed to what degree Nashville would prosper and that the Maxwell House would become the center of much of its social and political life, as well as attaining national and international fame.

Tired of the war and tired of the carpetbagger days, Nashvillians were ready to celebrate the opening of the Maxwell House on September 22, 1869, with a grand ball in honor of the Colonel's lady, the former Harriet Virginia Maxwell, for whom the hotel was named.

A description of the affair: The long-awaited opening was celebrated by a gala dinner (or supper, as it was called) in the enormous ballroom or dining room. Guests arrived through the main entrance on Cherry Street, the doors opening between eight fluted Corinthian columns. Guests stepped onto the black and white marble floor of the huge lobby with its two-storied height surrounded by the elegant balcony railing of the mezzanine. The high ceiling and the promenade balcony were supported by more Corinthian columns, and dark-stained mahogany arched doorways led to various hallways. The diner's destination and the pride of the grand hotel was the 150×53 foot dining room, opening off the balcony. It, too, was made elegant by six fluted columns at either end. The center of the ceiling was higher, with the tiered chandelier fashioned from cannon wheels. A huge ornate mirror decorated the mantel and fireplace opposite the dining room entrance.

The lobby was lined with tall, highly-polished mahogany counters, gleaming plate glass cabinets and polished brass fixtures.

Off the lobby was located telegraph and railroad ticket offices, shaving salons, a news depot, billiard rooms and a first class bar.

The original hotel contained more than two hundred rooms, parlors, dining rooms, ordinaries, promenades, corridors and the "grand rotunda".

Modern innovations for its day, the hotel offered bathrooms and water-closets on each floor. All bedrooms had a coal-fired grate. Steam powered passenger and baggage elevators and a public address system enabled the hotel management to keep in touch with all parts of the building. All kitchen fixtures needed in a first-rate establishment were installed. Everything for the convenience of the guests, as well as the workers at the hotel, was ordered with the "best materials and workmanship available".

Through the 102 years of its existence brilliant balls, banquets, Christmas parties, receptions and the rowdy hullabaloo of political rallies echoed from its grand ballroom.

The guest register of the Maxwell House read like a "Who's Who" of famous persons. At least eight Presidents enjoyed the hotel's hospitality: Rutherford B. Hayes, Andrew Johnson, Grover Cleveland, Benjamin Harrison, William McKinley, William Howard Taft, Woodrow Wilson, Theodore Roosevelt. Other distinguished guests included, William Jennings Bryan, Henry Ford, Thomas A. Edison, Annie Oakley and O. Henry.

During its 100 years of continuous operation at Fourth and Church, one could step off the bustle of busy streets into the Maxwell House lobby and tread the red carpets of an earlier world. Unconsciously, the pace slowed and voices softened. It was one of the few places that invited one to "slow down," advising that haste is useless, time less insistent, and cares less important. But all of this was to change. Nashville had enjoyed a beautiful but bitter cold Christmas Day in 1961. As night came on, news was flashed on radio and television that the historic Maxwell House was on fire.

The blaze had first been discovered by a guest, Miss Alice Earls, 19, of Chicago. Miss Earls said when she smelled the smoke she thought it could have been a cigarette burning in her room. When she opened

The Maxwell House fire in 1961.

the door she found a fourth floor hallway full of black acrid smoke. She quickly notified the switchboard operator who in turn summoned the fire department.

Because veteran firemen were aware of the dangerous possibilities, they were able to move into position in record time when the blaze was reported. Within a matter of minutes every available piece of equipment, including four hook and ladder companies, was on the scene.

Despite these efforts, within fifteen minutes a holocaust was born. The raging fire came as a surprise to everyone except city firemen. "Of course we didn't have a date set on it," said soot-smeared Fire Chief John Ragsdale, "But we knew by the law of averages it would catch up with us someday."

Urgent messages for off-duty firemen to report to the scene were blasted over radio, telephone and television.

Firemen armed with high pressure hose nozzles took up positions atop Burke and Company and poured steady streams of water on the spreading flames.

Meanwhile, inside the building an intense drama of man against unbeatable odds was being waged. Firemen were fighting the fire by going up to the top three floors on a rear stairway and ladders raised by fire trucks from the street.

A blast of red hot smoke belched from the fourth floor. A fireman shouted to his buddies, "Watch it! She's backdrafting!" The smoke and the red glow at the end of the corridor became more intense. A moment later a fireman rushed headlong through the door screaming, "Get more lines, get more lines in here!" It was all to no avail; clouds of orange-tinted smoke erupted through slits in the roof, engulfing the entire downtown area in a blinding fire-fed fog. The firemen fought stubbornly to bring the angry blaze under control. Outside could be seen a bright orange glow in windows along the rear of both the fourth and fifth floors. The glow was spreading rapidly along the Church Street side of the hotel.

At 10:30 p.m. flames broke through the 102-year-old hotel roof. Sparks showered the downtown area as flames shot fifty feet into the air. Within five minutes, two-thirds of the roof was a holocaust of leaping flames. Flames and sparks continued to shoot so high in the air they could be seen for miles. Billows of stifling gray smoke made it almost impossible to see the firemen working frantically from their tall ladders.

Like giant snakes, hundreds of yards of hoses coiled along Church Street and Fourth and Fifth Avenues. The firemen poured tons of water into the smoke-filled building. Although the firemen fought heroically and stubbornly with every piece of equipment available, the blaze roared out of control for six hours.

As plaster and wood began to fall in the lobby, all the city's firemen and equipment were called to retreat from the building.

At 10:45 Maurice Schutt, drill master for the fire department, said, "It looks like it's gone." With tears in his eyes, he said firemen had received orders to pull back. "We are going to have to let it go. That's our orders."

Each fireman was surrounded by a cloud of smoke as he left the lobby and walked into the street, disheartened and exhausted.

When it was thought the wall might collapse, policemen kept spectators at a safe distance. But the outside walls held. The flames ate down through the inner parts of the hotel.

Water damage alone ruined several businesses and restaurants on the ground floor and in its basement.

The Nashville Water Works Department estimated that 1,300,000 gallons of water had been used between 9:00 p.m. and midnight, most of it on the fire. The department said the reservoir dropped three inches during a period when it normally rises by six inches—a nine-inch variation.

As firemen fought heroically to bring the roaring fire under control, another traumatic drama was in progress—evacuation of the hotel guests. Firemen led elderly guests, many in their nightclothes, to safety. Other firemen systematically searched the building for guests who might still be in the building.

Mrs. Lucille Fraser was the night clerk and switchboard operator at the Maxwell House. Early that Christmas morning she helped her two sons open their gifts, little knowing that she would become a heroine before she saw them again. After receiving the fire call from Miss Earls, she immediately called the fire department and the hotel manager, then started notifying the sixty-five guests in the building to evacuate.

One guest, Mr. Silas P. Grady, 78, a retired state employee and permanent resident of Room 508, did not have a telephone in his room. About thirty minutes earlier he had returned to the hotel with an armful of gifts, greeted Mrs. Fraser, picked up his key and disappeared into the elevator. The Fire Chief reported Grady's room was searched before the flames went out of control. The restroom across the corridor was not searched because of the intense flames. "We just couldn't get in there," the Fire Chief quoted a fireman. His body was found later.

Twelve hours after the blaze was first discovered the weary firemen in ice-covered coats and boots and hats were still pumping tons of water on the spongy debris. The following day the Fire Marshal's office ordered the skeleton outside walls be torn down.

Thus the passing of a beloved landmark from the Nashville scene. Nothing—not even fire—can erase the Maxwell House's place in the lives of Nashvillians or in the history of a now modern city.

MOST BEAUTIFUL FIRE

I do not recall the date, but it was cold weather and I was working on Truck 1 from headquarters. At six in the evening, Truck 1 and Engine 2 arrived on the scene at the Exchange Building. The Exchange Club was a four-story brick building used for offices.

Forrest Escue was Acting Lieutenant on Truck 1. I was first in the first-floor entrance with Escue behind me. Smoke was visible, coming out the front door, when we arrived. The high arched doorway and open hall inside were rapidly filling with smoke, but we were still able to walk upright. About fifty feet inside, I found a door hot to touch and I could hear the fire roaring behind the door. I hollered to Escue and we went back outside. Meanwhile, Don Hobbs, driver of Engine 2, had hooked up to a plug in the alley on Church Street between Third and Fourth Avenues.

I went down the alley to the Carousel Bar, located in the rear basement of the Exchange Building. The Basement Bar was located in the front of the basement with an entrance from Church Street. As I entered the Carousel from the alley I heard Lieutenant "Humpy" Hahn call for an ax. That ax man was me. Humpy had a booster line charged and he told me to cut him a hole in the wall. We were in a restroom at the end of the bar. The wall was masonite on wood frame; one lick and I knocked a hole about the size of my head.

We could see the flames and hear the swoosh as the fire sucked in the air. Humpy told me to get Don hooked up. I told him Engine 2 was all ready to pump. He said to get him a 2½″ while he tried to hold the fire with the booster.

When I walked into the alley, I heard a sudden poof. I looked up to see every window on the first floor lit up with flames and belching smoke. I hurried to Don with Humpy's request for the 2½″ line. Don was already doing his job. By the time I reached Church Street I heard three more sounds—poof, poof, poof.

Each time I looked up, every window on that floor would light up simultaneously, as if a switch had been thrown. In a matter of a very few minutes the entire building was fully involved.

We poured water all night and were relieved by the day crew next morning. That evening, Truck 1 relieved the day crew. For two days and nights this continued until all hot spots were out.

Incidentally, this fire started in the room with the hot door. Painters had left open cans of paints and thinners. The fire was probably started by an undetected cigarette ash and accelerated by the painters' materials. The open stairwell helped the spread of the fire.

It was a tragic loss of a fine old building. But it was the most beautiful fire I can remember. It was almost like turning on the lights of a Christmas tree, the way those windows in each floor lit up.

Jim Cundiff - Retiree

THE FIRE AT THE FAIRGROUNDS

All the sights and sounds and smells of a perfect State Fair night were there on September 20, 1965. It was Monday, always designated in Tennessee as Children's Day.

In the midway at the bottom of the hill were the vivid, flashing neon lights, the ear-splitting music, the barkers urging the onlookers in to see the fat lady, the man with a human head and a snake body and other typical carnival sights. The air was filled with the pungent smell of onions frying with hamburgers, the greasy hot smoke from corn dogs cooking, mixed with the sticky, sweet smell of cotton candy. Thousands of people were enjoying the chills and thrills that were a Fall tradition for Tennesseans.

At 10:30 p.m. one word changed this carefree frolic to hellish chaos. One woman looked up the hill and screamed, "Fire!"

Courtesy of Metro Nashville Archives

Fire at fairgrounds on September 20, 1965.

The music stopped, the barkers threw down their microphones, the food vendors jerked to attention. Everybody started screaming: "Fire!" "Fire!" Almost as a body of one, the thousands of people in the midway started running toward the fire that was now leaping two hundred feet in the air.

Up on the hill hundreds of people had gathered in the grandstand adjacent to the Woman's Building for the free acts and the fireworks display that followed. It was that one word, "Fire", that changed this tranquil scene to an instant riot.

Everybody started screaming, "Fire!" They were running, jumping, clawing—anything to get away from the fire. Crowds running away from the fire met the crowds running to the fire, causing mass confusion. Mothers and fathers were grabbing children. Younger people were trying to help older people escape. Souvenir vendors were snatching their money boxes and armloads of their wares. Food vendors were grabbing whatever they could. Looters, taking advantage of the situation, made off with anything that was loose as they ran for safety.

Everybody realized they had precious little time to save themselves.

Within two minutes the fire was licking up the walls of the ramshackled Merchant's building. The dirty floors, unpainted walls covered with gaudy paper streamers, and the garish advertising signs added fuel to the flames that quickly spread to the Woman's Building.

The fire company stationed at the Fairgrounds was immediately on the scene. Instantly, they radioed for all help available.

Fire engines, fire cars, ambulances and police cars, all with flashing lights and blaring sirens, shattered the night as they rushed to the fairgrounds. On arriving at the entrance they were blocked by the swarming, panicked mob and the carloads of people attempting to leave the area. Policemen were struggling to assist pedestrians, route tangled traffic out of the area and clear the way for emergency equipment to get to the fire.

At the Merchant's Building, terrorized people began jumping forty feet from the third-story windows. Several people with clothing blazing rushed out of the building. Young girls with burning embers in their hair had to be forcibly stopped. Hysterical children were separated from their parents.

A tearful man ran from the building. He reported that there were some people trapped in the building. Rescuers rushed in to save them. Pandemonium was everywhere.

Phillips-Robinson Funeral Home had two ambulances around the clock during the fair. An ambulance attendant reported, "I took off up toward the fire but we could not turn the ambulance around after we got there because of the crowds. A big wind came up and the fire got hot. I'm telling you that was the hottest fire I've ever seen. I've been in all kinds of disasters but this is the worst ever."

It began to sound like a battle zone as the roaring wind blew flames to the parking lot. Several cars caught on fire, causing their gas tanks to explode. Electrical transformers started blowing up and "hot" electrical wires dropped to the ground, adding more danger to the situation.

Flames ate through the grandstand, leaving thick steel girders bent like hairpins. Automobiles and trailers parked by the Woman's Building were bent and gutted.

All of a sudden the whole roof of the Woman's Building collapsed. Flames shot hundreds of feet into the air. Sparks and burning embers showered the entire fairground area. The wind caught the sparks and carried them as far as Fourth and Rains Avenue, one mile away. It struck horror in the hearts of all the residents in the area. It was reminiscent of the horrible East Nashville fire of 1916 as they, too, used hosepipes to wet down their roofs.

In the minds of many people, the Woman's Building was the State Fair at its best. The historic brick structure had a jaunty Victorian air.

Coveted ribbons were awarded for the best of everything Tennessee produced. In the food area the fair goer could see the prize-winning vegetables, beans, pumpkins, squash, potatoes and corn. Canned fruits and vegetables lined the walls. Prize-winning breads, cakes, and pies were on display. One could always smell and sample Tennessee's finest country ham.

One large area was devoted to all types of needlework—such as crocheted doilies, antique clothing and beautiful quilts. Another area was devoted to art—professional, primitive, and school children's pictures drew an appreciative audience.

Collectibles of all types filled display cases. The flower show competition was the highlight for statewide flower lovers. Davidson County Home Demonstration Clubs worked all year to complete more than a score of educational booths that delighted the homemakers.

In the center of the main floor rotunda was a reflecting pool with water lilies and goldfish. Fair goers gathered here for musical programs and 4-H Club fashion show competition.

The Woman's Building had closed at 9:00 p.m.—an hour and a half before the fire started. Had it been filled with people, it is feared that many would have lost their lives.

All the beautiful, priceless articles were gone— completely destroyed. Gone also was a wonderful tradition that would never again be recaptured.

Robert Hawkins, who was in charge of the Tennessee Game and Fish Commission display tent located beside the Woman's Building, gave this comment: "As soon as I saw the fire over there we immediately began to move all the animals out. We loaded the deer into a station wagon and then secured the poisonous snakes. Policemen guarded the snake cages with shotguns. We moved our bears in their cages down by the stockyards until the stockmen threatened to have them shot because they were making the cattle nervous. Some of the aquariums broke and some of the fish cooked while others swam around in the boiling water as if nothing had happened."

As the fire raged out of control, the livestock exhibit of 1,456 animals, valued at one million dollars and located only three hundred feet from the origin of the fire, had to be evacuated. Working in the light of the fire, the exhibitors led their animals from their stalls, down the hill to the Brandsford Avenue side of the fairgrounds. It was a long hectic night for them because of the excessive heat and the deafening noise. The bears from the wildlife exhibit made the cattle restless.

There were no casualties among the livestock, but the valuable chicken and rabbit exhibit was destroyed.

The 4-H Club displays with valuable farm machinery and the Al Ameniah motorcycles and trophies were burned.

At midnight police moved everyone except emergency personnel out of the area as the fire threatened the principal gas main running beneath the fairgrounds. Firemen said the blaze had not reached the gas line but the heat was so intense they feared the main might explode.

As word spread throughout the city, switchboards of the fire, police and news offices were flooded with calls from concerned residents who had relatives at the fair.

When word of the fire was first received at General Hospital, interns and nurses in the emergency ward were put on alert for casualties that were certain to arrive. Dr. W. J. Core, Metro medical examiner, arrived at General Hospital shortly after 1:00 a.m. to stand by in case fatalities were reported.

As firemen gained control of the blaze, these buildings lay in ruins: the Woman's Building, the grandstand, numerous restaurants around the buildings, the 4-H Building, the Administration Building.

Eighteen persons, including ten firemen, were treated at local hospitals. It was a miracle that there were no fatalities.

The following morning, grimy-faced rescue workers searched for bodies in the still smoking acres of charred rubble left in the wake of the ravaging windswept fire. The grim search was launched shortly after daybreak, seven hours after the fairgrounds burned like a bombed out city.

The firemen had put up a valiant fight against unbeatable odds. Low water pressure hampered their efforts to get enough water to the wooden Merchant's Building, filled with inflammable material which caused the fire to spread quickly. The mob of hysterical people hampered the arrival of fire-fighting men and equipment. The strong wind had fanned the flames into unbearable heat.

As Metro Fire Chief Russell Campbell surveyed the smoldering pile of rubble that was once the Woman's Building he said, "It's the biggest single structure fire in our history. We had ten firemen sent to General Hospital overcome with smoke. It got so hot we had to move an engine back from the building to keep it from melting. This city should be very proud of my men who risked their lives."

The fire did ten to twelve million dollars damage that night. Also destroyed was a grand tradition that has never been recaptured.

STATE FAIRGROUNDS FIRE

I was assigned to host the Fire Prevention Booth at the State Fair in September, 1965.

The first day was uneventful and the crowd in the Woman's Building was beginning to dwindle as the booths were secured for the night. The time was eight o'clock.

The Woman's Building was a two-story wood building. It was joined at the second level with the wooden grandstand. The street ran between the grandstand and the Woman's Building, forming a tunnel under the second level. On each side of the street were numerous booths serving food and selling souvenirs. The aroma of food was everywhere. The crowd increased, anticipating the grandstand acts and the fireworks.

I left the Woman's Building and met another fireman, John W. Owen. We were watching the crowd come and go as we stood talking on the steps behind the grandstand. We decided to go through the midway before going home for the night.

About 8:30, John and I crossed the old dirt race track and were entering the midway when I spotted a couple of friends. One of them remarked, "Jim, I see a lot of sparks."

I turned around to see sparks flying thirty to forty feet in the air around the Woman's Building. I ran back across the race track and pulled the alarm box at the Fair Park. Looking up, I could see that the tunnel was filled with fire and smoke. Then I heard Engine 6B from the other side of the tunnel hit the siren. Engine 6B was a "make up" company stationed at the fairgrounds. I don't believe we had radios at that time, other than in the Chief's cars. Engines 5 and 9, Truck 2 from Fourth and McGavock, Engine 8 from 8th Avenue South, Engine 15 from 12th and Caldwell, Engine 6 from 2nd Avenue South and Engine 12 from

Warf Avenue all responded to the fire box alarm. Later the Chief called in Engine 17 from 17th and Division. I do remember vividly how I tried to stop people from stampeding out of the Fair Park. I told them not to run, but to hurry out of the way, to watch for the fire trucks and to leave the area. I remember stopping a running fifteen-year-old girl with long red hair. Just as I stopped her, a big piece of burning debris the size of a half dollar fell into her hair. I held her long enough to get the smoldering wood from her hair. I shudder to think of what might have happened if she'd kept on running.

By now the grandstand was on fire. Flames were shooting hundreds of feet in the air. Both buildings were fully involved. There was no way to extinguish the fire; we had to try to keep it from spreading. I was busy helping lay out additional lines and relieving some of the others. I saw that it was going to be a long night for everyone. Noon the second day: Grandstand lies smoldering among the twisted steel girders, standing guard over what was left of many years of memories. The Woman's Building is nothing but ashes. At the bottom of the hill, the horse-show building has many windows broken from the intense heat from the previous night. Horse-barn stalls nearby are scorched and empty.

Luckily, no one was seriously injured. No stock or horses lost—just buildings and their contents. One death was reported earlier but it turned out to be a false alarm. There were casualties among the rabbits and chickens underneath the south end of the grandstand.

Since the fire the fair has lost some of its glamour and appeal, at least for me. No more dirt track with the horse racing; on Saturdays the Offenhauser Race cars on the 1-1/10 mile dirt track, throwing dirt on the crowds as they lined the outside curves of the track; no crowds on Morris Hill to watch the racing and the fireworks. With the fire came the ending of an era.

Jim Cundiff - Retiree

RADIO COMMUNICATION REPLACED

In September of 1956 the radio communication system for the fire department was put into service.

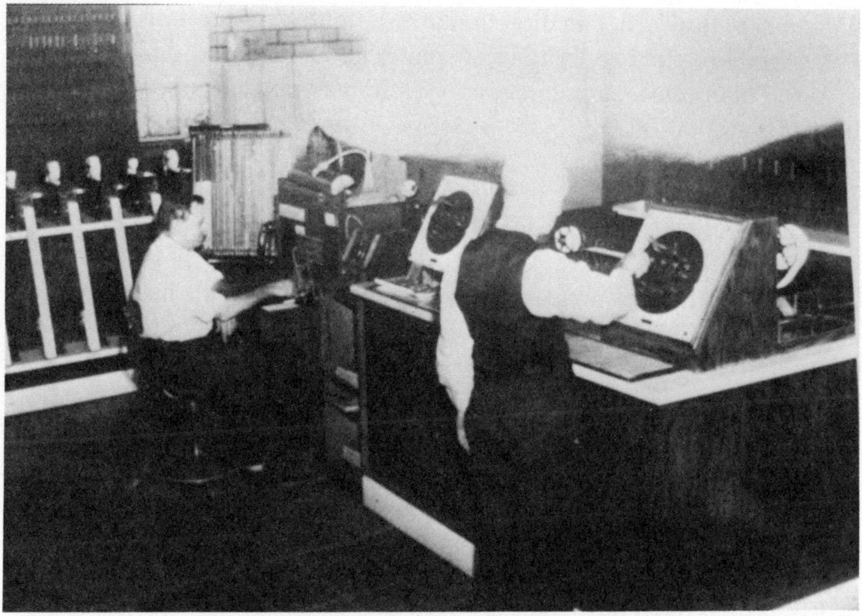

Alarm Station, 312 25th Avenue N., early 1940s.

It consisted of 42 specially designed mobile two-way radio sets in all Fire Department equipment, plus 11 two-way mobile radio sets for the fire chiefs and staff offices. Also included were the portable control stations, a main base station operating at 500 watts and six "handy walkie" units. General equipment included a 100-foot tower at the base station and a 10,000 watt gasoline emergency generator.

Fire Chief H. D. Demonbreun, Mayor Ben West, and Robert Cassetty, superintendent of the fire alarm division, predicted increased efficiency for the department under scope of fire prevention inspections of commercial, industrial and multi-family dwellings and a possible reduction in fire insurance rates as a direct result of the new system.

Using the new two-way communication system, headquarters and department chiefs are in constant touch with each company.

With the radio system, firemen can be warned of the character of the blaze en route to the fire.

In the event of a disaster, men and equipment can be routed, recalled and directed from master control stations with a minimum loss of time and speed. The small "handy talkie" radios plus public address announcer allow efficient direction of fire fighting by staff officers and "tie together" the fire-fighting operation into a fully coordinated team effort. This radio communication system was put into operation at a cost of $40,000.00.

A HISTORY OF THE NASHVILLE FIRE BUFFS

The first formal example of "Fire Buffing" in Nashville had its roots in 1952 when Chief John D. Ragsdale organized the Nashville Fire Buff Club.

The first Buff truck was the "Ben West Disaster Unit," a used forty-passenger city bus obtained from the Nashville Transit Company. Named for the former Mayor of Nashville, the bus was reworked by the fire department shops. The unit not only had the usual equipment like coffee pots, water tanks, and a place to prepare food but also was a mini-emergency room. The unit carried drugs and other medical equipment as well as having a physician in attendance at all multiple alarm fires. The membership in this club reached 35 and played a major role in serving the old city fire department.

Declining interest through the years gave way to the next generation to show an interest in buffing. This was the Music City Pompier Club formed in July of 1979. This group of 15 included many names that would later become very familiar in the Nashville Fire Department. Members included Ronnie Roller, presently Deputy Chief, Ambulance Division; Mark Cohen, Lieutenant, Communications; Ken Nelms, District Chief, Firefighting; and Robert Brooks, District Chief, Ambulance Division.

Bill Ketchum proved to be the workhorse that held the hope for firefighters that someone would be there when they needed help on the scene. Often, working on the scene by himself or staying all night and then going to work, Bill and his son did it all. If it needed to be done, Bill was ready. The Fire Department gave him an old ambulance. He was the Fire Buff Club for several years.

On December 14, 1987, at Engine Company 12 a group of nine became the charter members of the Buff Club that would become known as the Box 55 Association. The charter roster included Bill Ketchum, Jim Gregg, Lon Tenson, Bob Bomar, H. P. Harris, Scott Harris, Buck Dozier, Richard Perkins and Wayne Fisher.

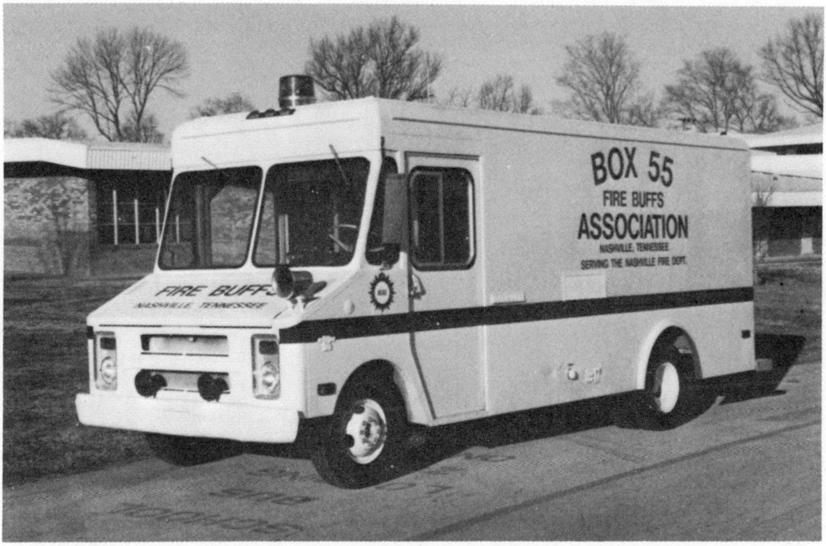

Van used by Fire Buffs.

With nothing except the determination and goal to provide service the firefighters of Nashville needed, the club began to grow not only in number but in terms of service. Councilman Buck Dozier, a member himself, sponsored a bill that made us an auxiliary unit of the Fire Department. Officially termed The Auxiliary Emergency Assistance Organization (AEAO), this opened up new ways we would be able to serve. We were given the use of old salvage 2 when the new salvage truck was put in service. The department shops reworked the truck, adding a serving window on the side, a generator, lighting, and other items to provide better service on the scene. The club now has officers and plans to be around for many years to come. Our thanks go out to the men and women of the Nashville Fire Department, to Chief Martin Coleman and Chief Elza Smith and the people down at the shop, to B. R. Hall and Firefighters Local 763, and countless others.

Our motto is "Serving the Nashville Fire Department." That will always be our goal for this brave group of men and women.

Richard Perkins and Bill Ketchum

THIS IS THE FIRE DEPARTMENT

CHIEF
- leaps tall buildings with a single bound
- is more powerful than a locomotive
- is faster than a speeding bullet
- talks to God

DEPUTY CHIEF
- leaps over short buildings with a single bound
- is more powerful than a switch engine
- is just as fast as a speeding bullet

DISTRICT CHIEF
- leaps short buildings with a running start and favorable winds
- is almost as powerful as a switch engine
- is faster than a speeding B.B.
- walks on water in indoor swimming pools
- talks to God if special request is granted

LIEUTENANT
- barely clears short buildings
- loses tug-of-war with switch engine
- can fire a speeding bullet
- swims well
- is occasionally addressed by God

DRIVER
- makes high marks when trying to leap buildings
- is run over by locomotive
- can sometimes handle hose without inflicting self injury
- dog paddles
- talks to animals

ASSISTANT DRIVER
- runs into buildings
- recognizes locomotives two out of three times
- is not issued fire extinguisher under any circumstances
- can stay afloat if properly instructed in use of life jacket

FIREMAN
- falls over doorstep when trying to enter building
- says "Look at the choo choo"
- wets himself with a water pistol
- plays in mud puddles
- talks to himself

CAPTAIN
- lifts buildings and walks under them
- kicks locomotives off the tracks
- catches speeding bullets in his teeth and eats them
- freezes water at a single glance
- because.......he is GOD

Anonymous

Firemen dressed as clowns participated in huge Fire Prevention parades in the 1950s and 1960s.

Recruit graduating class of 1955.

2 Stories by Firemen

STUCK!

I was Assistant Chief at the time this happened. Lieutenant Richard Kimbro was my driver. We got a call for my car and two companies to respond to a house off Belmont Boulevard. The operator said, "Someone is stuck in a bathtub." Having never received a call exactly like that, we discussed the possibility that it might be a baby drowning in a tub of water.

Arriving first on the scene, we met a middle-aged lady standing on the porch. She said, "My mother is stuck in the bathtub and I can't get her out."

I thought, "We ought to be able to handle this." I gave all the other companies a signal X.

We went in to find a very large, elderly lady covered up with a sheet, sitting in the tub. The tub was enclosed with glass shower doors, which complicated the problem. After looking the situation over, I said, "Now, lady, I'm an old married man so don't you get upset. I'm going to have to get over in the bathtub with you and I'll have to put my arms around you. You are going to have to hold on to me, but I'll get you out of here."

Facing her, I took hold of her and she grabbed hold of me. I wiggled her back and forth. She became unstuck. It took some very careful maneuvering to get her to stand up and the two of us out of those shower doors, all the time trying to keep the sheet in place.

Finally we got out. Both of us were out of breath and exhausted. I practically had to carry her into her bedroom, where I left her sitting on the bed.

We checked back in service, hoping that the next call would be a regular fire call.

Will P. Ackerman - Retired Chief

OH! NOT NOW!

When I was Captain at 19's it was a triple company so there were a lot of men at the hall. We had just come in after a run. We were all in the office laughing and drinking coffee, when a very pregnant lady came rushing in and collapsed in a chair. She announced, "I'm having a baby! Quick! Somebody call my husband, Ronnie Neeley. Tell him to come over here now!''

You should have seen how quick those firemen scattered. She and I were left in the office alone.

She had a pain and groaned.

I pleaded, ''Oh! No! Lady, not now!''

She had another pain and another groan.

I pleaded even harder, ''Please, lady, wait just a minute! Ronnie is on his way.''

After what seemed an eternity to me, Ronnie arrived and took her to the hospital.

I saw her at the ''Cabbage Cookoff'' this spring. She told me that that baby is now twenty-three. He is trying to get on the Fire Department. It brought back memories of how close he came to being born at the firehall.

Bill Stoutt - Retiree

A TALE TWICE TOLD

[Editor's note - I called Mrs. Ronnie Neeley to ask permission to print Bill Stoutt's story, **Oh! Not Now!**, because it was so personal. Her response, amid peals of laughter, ''Well, it's funny now but it wasn't funny then. Captain Stoutt is too kind. Let me tell you how it really happened.'' Her version follows.]

I can tell you exactly when it happened—October 20, 1965, at 2:30 in the afternoon. I had been to town to pick up some last minute things because it was just a few days before I was due to go to the hospital for the birth of our baby.

As I drove out Charlotte I felt my first labor pain. All I could think of was, *Get to the firehall. They will locate my husband, Ronnie.* He was the driver for Chief Tidwell.

I wheeled into 19's, ran up the ramp and into the office. I collapsed into the nearest chair.

All of the firemen had just gotten in from a run and had gathered in the office and were drinking coffee.

I blurted out, ''I'm having a baby! Get Ronnie Neeley on the radio!''

Pure pandemonium broke out. Firemen scattered every which way. One fireman stumbled over an office chair getting to the telephone, hit his head on the corner of the desk and started bleeding. After all the confusion died down, Captain Bill Stoutt and I were the only ones left in the office. I could hear the radio keep blaring, over and over, "Emergency at 19's. Ronnie Neeley call Engine Company 19."

Captain Stoutt was so kind. He pleaded, "Mrs. Neeley, please don't have this baby here at the firehall. Let us take you on to the hospital. They'll soon find Ronnie. He'll come straight to the hospital. I'll even stay in the labor room with you until he gets there."

A pain would hit me and I would have to groan.

Captain Stoutt was getting more nervous all the time. "Please, Lady, not now!"

The telephone rang several times. I heard him shout, "We are having a baby!" and hang up.

He got so upset he called the operator and told him, "Tell everybody to quit calling over here. We are having a baby and I don't have time to answer the phone!"

As it turned out, Chief Tidwell and Ronnie had checked out of service for a planning meeting at the Training Academy. They had their radio turned off.

Ronnie just happened to pass by the Academy's radio and heard "Emergency at 19's." He called the operator to find out what was going on. The operator shouted, "You are having a baby at 19's."

In a few minutes the red Chief's car, with the red lights flashing, screeched to a halt on the ramp at 19's. Chief Tidwell jumped out. The firemen tenderly helped me into the car. I heard Chief Tidwell holler, "Don't you all have a wreck!" as we sped away.

Less than ten minutes later we had our baby boy.

The next day I received a huge bouquet of flowers from the firemen. The card read, "Congratulations! Thanks for not having the baby at the firehall. Engine Company 19."

At the *Daigo Cabbage Cookoff* this spring Ronnie said, "Honey, do you remember this gentleman, Captain Bill Stoutt?"

"Oh, Captain Stoutt, I'll never forget how kind you were to me. Now, I'd like for you to meet that baby!" as I pointed with pride to our six-foot son.

Mrs. Ronnie Neeley - Wife of Captain Ronnie Neeley

BE PATIENT YOUR TIME WILL COME

Red Meadows and I were always into something. Some of our schemes were done on the spur of the moment; these seemed to come natural. Other schemes took a little forethought. This particular one took a little planning and a little waiting for just the right timing. We patiently waited; our perfect time came.

We worked with two "scaredy cats." They were afraid of everything and stuck pretty close to each other for protection.

At 17's our furnace was in the basement. It was supposed to operate automatically, but sometimes it would malfunction. When it did somebody had to go down the steps to trip the switch.

This malfunction come one late afternoon when we were waiting for our relief men. The two scaredy cats jumped up and ran down the steps to trip the switch. At last our perfect time had come. I happened to have some little firecrackers that I had saved for just this opportunity. When we heard them trip the switch I lit a firecracker and dropped it in the register. They thought the furnace had blown up.

They like to have killed each other getting up the steps. We almost killed ourselves laughing.

D. White - Retiree

A REAL COOL FIREMAN

I was over at Engine Company 4. I hadn't been on the Fire Department very long. Our engine was old, one of those you rode standing on the back. It didn't have a top on the front.

We got a box call to 12th and Russell. When the bell rang I was in the locker room, dressed only in my pants and a T-shirt. I managed to get one boot on, grabbed my other boot and my coat. It was ten degrees and snowing. We drove real fast all the way from 5th and Woodland to 12th and Russell. It was definitely not T-shirt weather—I was frozen.

All of this for a false alarm! As we returned to the hall the engineer drove real slow. He wanted to teach his rookie fireman how cold he really could get.

From then on I made sure I put my coat on first!

Wayne Vick - Engine 1

SHE GOT ALL MAD

A young boy came to work in the ambulance division. We called him "Muscles"—he was so skinny. He told us he was getting married soon.

We all got together and decided that we needed to give him a unique wedding gift, something he and his wife would remember and could even brag about years later. It took two days and one night of concentration and planning to decide what the gift would be.

In preparation I got a bottle of tincture of violet, that purple stuff that won't wash off for days. We knew it was safe to use because they used to put it on kid's scrapes and scratches.

Our plan was that when they went on their honeymoon to the beach in Florida, people should be able to identify him as being from Nashville, Tennessee.

The time came for us to deliver his gift. Several firemen held him down while the rest of us painted the Metro Nashville City seal on his chest, with a sunburst and other symbols. We even tried to paint the Indian, although he struggled hard.

When he came back from his honeymoon, he told us his wife got all mad about that. She wouldn't let him take his shirt off.

George Owen - Headquarters

MY DOG STORY

We have all heard stories of how firemen are called to rescue a puppy. After the job is done a very cute little blonde looks up with her beautiful blue eyes and says, "Thank you Mr. Fireman for rescuing my puppy. I'll always love firemen. You are so brave, strong, and fearless."

Believe me this is not the way it went when I was sent on a "dog call."

We received a non-emergency fire call to go down on Charlotte Avenue to a tiny church building. Captain Les Young said to the driver, Earl Percy Brown, "Come on, and we'll take Gordon."

I guess I was chosen because I was a new recruit. We were met at the door by a very small "Lady Preacher." She said, "There is a terrible odor in here. We can't have church services until something is done."

We cut a hole in the floor and located the smell. A big dog had breathed his last breath under the floor. Captain Young told me to go get a mask and a scoop off the truck, drop down about 5 or 6 feet, scoop up the dog and put it in a box, then hand it up over my head to Earl Percy. It was not a "nice, clean, happy, rescue the puppy situation." I had to turn my head and just aim the scoop in the direction of the remains.

The "Lady Preacher" was so fascinated by the horrible sight and smell she kept running up to the hole and looking down and running out to the porch and stomping her foot and gagging. Then she would come running back to the hole for another look. After she had made several trips, we got the job done.

I had the best part of the rescue operation because I had a mask and the others did not.

Gordon Hall - Engine 1

TOO LATE

My name is Henry Booker. I am a Paramedic in the Metro Nashville Fire Department. I came on in 1972. I have been a Paramedic since 1978.

Right after I graduated from Paramedic school, I was put to work on one of the busiest units in the city, Medic 18. On the first shift at 7 p.m., we got a call to go to North 1st Street behind Truck Stops of America. There were three disabled construction workers located approximately 1,000 feet into a drainage system that ran underground across the city. The opening was six feet in diameter and thirty feet down in the ground.

We were dispatched with Medic 5 which was located downtown. Just before we pulled up on the scene, a Paramedic named Eddie Montgomery got out of Medic 5, climbed down in the hole and went inside the drainage system. Some of the construction workers had escaped. They told us that two more of their friends were down there.

Montgomery went in with an oxygen cylinder. We never heard from him again. Five hours later we got his body out, along with the two construction workers.

The problem was that the construction workers, who had been overcome by methane, were so far back in the drainage system that the oxygen ran out on the breathing tanks. The air lasted for only twenty or twenty-five minutes. By the time the warning alarm went off and the person realized he was low on oxygen, he would be so far away that he could not get back to the opening.

A total of 19 firemen and EMTs went to the hospital that night with respiratory problems. Chief Red Davis was in charge. He ordered that no one was to enter the tunnel until they could find a way to guarantee their safety. He decided to go up the street about three blocks, drill a hole in the street, and ventilate the tunnel to get the poisonous gas out. After this was done, we got into the drainage system. We found Eddie and the two construction workers but it was too late to save them.

I remember doing CPR on Eddie Montgomery all the way to General Hospital. He had been in there too long. Everyone knew it was too late, but we still tried to do CPR.

Of all the years I've been a Paramedic, that was one of the worst situations I've ever been in—to have to do CPR on my friend knowing that it was too late.

Henry Booker - Rescue I

A GAS MILEAGE PROBLEM

The story I have concerns W______. He was an engineer on Engine 9 in the "Bottoms" in the early '60s. W______ was the type of person who was easy to "get to" and we took every opportunity to do that. In 1965 he bought a foreign car. I believe it was a Fiat. We gave him the devil about it. All the rest of us were driving American vehicles. He told us about what good mileage he was going to get. After he'd driven it a week or two, he told us that he was getting thirty-some miles to the gallon. We thought we'd help him do a little better than that. At that time we were on a rotating shift—working every other day for 12 hours. For about two weeks, one of us would add a gallon of gas to W______'s car. He began to brag that his gas mileage was 78 miles to the gallon!

We decided that it was time to bring him back down to earth a bit. Then each one of us would siphon a gallon of gas out of the car.

After a week or so, W______ checked his gas mileage and found he was getting only 12 miles to the gallon. It really upset him, of course.

We helped him on, telling him that we wouldn't have a car like that—getting 78 miles to the gallon for awhile, then getting only 12. We advised him to take it back to the dealer.

After a little coaxing, W______ decided that was the only fair thing to do. He went out to the Fiat dealer on Broad Street and raised sand with them. An hour later, he came back, as mad as he could ever be. We encouraged him that he was doing the right thing. We told him to go back to the dealer and demand his money back. We got him tuned up to go again. When he'd been gone about 30 minutes, we received a call from the Service Manager at the car dealer. They had finally figured out what we had done. They told us we'd better get the man back down there to the hall because they were fixing to have him arrested. At last we had to admit what we'd done. We sent a patrol car out there to escort W ____ back to the hall before we all got into trouble.

Terry Nickens, Captain – on the Fire Department 25 years

ROUND AND ROUND SHE GOES

I was Captain at 3's. I was on a late night watch alone. All of a sudden a very large lady came running in the door. She was sweating, all out of breath, and very upset.

"Call the police. There's a carload of young hoodlums trying to pick me up."

I told her she could use the phone to call the police if she wanted to. She did. After she hung up, she started around the table toward me. I started cautiously moving on around to the other side of the table.

She said, "You know my husband left me."

"No, I didn't know that."

Then she said, "Well, you know I just got out of Middle Tennessee (Mental Hospital) didn't you?"

I sure did not, but it didn't make me feel any better to hear it now. We made about three rounds circling the table. FINALLY, the police got there. I went out to their car and told them what had happened.

They said, "She's not getting in this car."

The policemen told her, "Lady you just walk on home and we'll follow you and protect you."

Last I saw she was walking down the street. The patrol car was following along real slow behind her.

Bill Stoutt - Retiree

I'LL GET YOU

Hilderbrand and Weakly were my buddies. They would gang up and pick on me. I had to figure out some ways to get back at them.

Before I went to bed one night, I heard them sniggering and knew they had been up to something. When I crawled into bed, I found out they had short-sheeted me. So I waited until they both went on a two-hour watch, then I got up and went to the kitchen to get some crackers. I crumbled the crackers up real fine and spread them out, all over their sheets. Then I pulled the spreads back in place.

I lost two good hours of sleep just waiting to hear them cussing me. It was worth the wait. I sniggered, too, as I watched them turning over their mattresses so they could sleep on the bottom side!

D. White - Retiree

MY FIRST FIRE CALL

Many changes have been made in the fire department since my appointment on September 15, 1955. One major change was in the method of training recruits.

On my first day, we were informed that school would be in session from 8:00 a.m. until noon, after which time half of the class would report to work with an assigned company until shift-change at 5:00 p.m. The other half would report at 5:00 p.m. to work until the next morning.

I reported to Captain John Alvin Cox at Engine 4, then located at 5th and Woodland Street, to work the night shift. We hadn't been issued our fire clothes, but Captain Cox told George Underhill to borrow from the day shift some fire clothes that came closest to fitting me.

He was showing me where to ride on the back of the Oren Pumper when suddenly the main line jingled, signaling that I was about to make my first fire call. By the time I got into my boots and fire coat, the engine was ready to go out the door. I jumped on the back of the engine, in the center as I'd been instructed. Hurriedly I put on the helmet, which was a bit too large. I didn't have time to adjust the chin strap before we were speeding down Woodland.

I held onto the cross bar with all my might, standing on my toes as I'd been told to do. I couldn't imagine standing up and riding so fast! I thought surely Engineer Jimmy Love must have his foot all the way to the floor. When he slowed down to turn into Third Street, I decided I might be able to tighten the chin strap on my helmet, but before I could turn one hand loose we were flying again.

As we topped the Old Meridian Street viaduct, we hit a bump and my feet flew out from under me. The wind caught my helmet, almost jerkng it off my head. I opened my mouth to try to clamp onto the strap; the wind billowing into my mouth made me feel like a balloon blowing up.

Finally, we stopped at Lischey Avenue and Mile End, only to discover it was a false alarm. But I felt grateful, because I knew I wouldn't have been any help — the way my legs were shaking.

I've made many fire calls in the last thirty-four years, but that first fire call is the one I will never forget.

Gordon Hall - Engine 1

I COULDN'T HAVE DONE IT

We took the Captain's cigar and soaked it in hot pepper juice. We put it on the radiator to dry out. When he came back off a run, he got the cigar and took a big puff that sent him all the way to his knees. He threatened me. I pleaded innocent, "Cap, I couldn't have done that. I was on the run with you."

He said, "You did it! I know you did it! I don't know how you did it, but you did it!"

George Owen - Headquarters

A MEMOIR

I joined what was the old Nashville City Fire Department on August 8, 1940, and worked that very night at Fire Headquarters on Ladder Company 1.

The regular firemen on duty that night initiated me by removing my night-pants and boots from beside my bed, replacing them with a pair of pants with the legs cut off way above the knees and a pair of very short boots. Without my knowledge, of course.

They had also wired my bed with small electric wires. Then they waited until I climbed into bed and was almost asleep. Some joker then gave me a couple shots of electricity to my rear, and at the same time pushed the fire alarm button.

Well sir, between that electricity popping me and the fire alarm bell going off, I was scared half to death. But I did manage to slide down the pole, thinking that I was also going to a fire.

Instead, as I landed on the floor, there stood all the firemen laughing at how foolish I looked in those short pants and boots!

"Brother" Clive Gibbs

TOO OLD AND TOO SLOW

I remember my granddaddy, A.W. Bebout, telling me a story about his old days as a fireman. Once, when their regular truck was in the shop, they had to use an old, very slow-running reserve truck.

They were out making a run up Charlotte Pike, and the fire truck could hardly pull the hill. A small boy on his bicycle came riding along beside the truck. "Hey, where ya goin'!" he shouted.

The captain gave his men a wry look. "Well, if we can't beat a bicycle," he declared, "just turn off the siren!"

Mike Russell - Rescue 1

BUGGED

D. White, a retiree from the Fire Department, tells how one of his fellow firemen was scared to death of a bug. "Of an evening, when ol' Dean was ready to go home, he'd always hang his clothes over the locker door and go in to take a shower. We all knew how scared he was of bugs, and since Cy Gordon had brought me a handful of cockleburrs to keep in my locker in case some opportunity for a practical joke came along, we decided to use them on J.I. Dean.

"While Dean was in the shower, I placed three cockleburrs in the crotch of his underwear. When Dean began to dress, those cockleburrs hit him and he went to squalling and kicking.

"I swear he tore that underwear all to pieces trying to get it off!"

D. White - Retiree

THEY TAUGHT ME TO BE ON TIME

The first day on the job, I was late. My daddy, a fireman too, had been working at "The Bottoms" and he stopped by the firehall to see how his little boy was doing that first day on the job. When he found out I was late, that I'd overslept, he decided to wait until I showed up.

All the firemen had known me since I was a kid, for I'd grown up around the firehalls. Of course, no one called me; they just waited to see how late I'd be. At last, when I woke up and saw what time it was and how late I was, I panicked. I threw on my clothes and flew into the firehall—and there sat my Daddy. I turned red as a beet and just sat down.

The Captain said: "Well, Vick, what do you think I ought to do? First day on the job and he's late."

"I think you ought to take him to Headquarters, to the Chief, and ask for him to be fired," Daddy said.

The Captain was merciful, although I still arrived late at times. I tried not to let it happen, but being new on the job and getting up so early—I just wasn't used to it. And still they wouldn't call me; they'd all just wait so they could "get me."

Sometimes if I was late, I'd park in the alley across the street and try to sneak in. But my fellow firemen had scouts all over the place, watching for me.

I could always tell when they were going to get me, because it would be dark inside. The doors were all shut and no one would be stirring. Then the minute I'd start to creep in, every siren would blow. Bells would ring. Doors would be flung open and everyone would cheer and holler. All I could do was to stand there and feel like a fool.

That all happened to me several times. I finally got to where I could get up on time.

Wayne Vick - Engine 1

SHAKE IT HARD

A telephone booth was conveniently located on 4th Avenue across from the fire hall in "The Bottoms." We could talk on the fire hall phone while looking across the street to see whatever might be happening in the phone booth. There was usually a lot of activity, because it was often used by the winos who frequented that area.

One day a fireman saw a wino going into the phone booth. Quickly, he rang the familiar number and the wino answered. The fireman said, "This is the telephone company. Did you just make a long distance call from this booth?" Before the wino could answer, the fireman added, "You have a refund coming back to you. Please check the slot."

The wino checked and replied, "There ain't nothing there."

"The money must be stuck," the fireman said. "You'll have to go outside and shake the phone booth very hard. Then the coins will come back."

The man hurried outside to shake the phone booth so hard that a policeman, driving by, thought he was trying to tear it up and hauled him off to jail.

Wayne Vick - Engine 1

"YOU LOUSY FIREMEN!"

One payday we decided to pull a prank on our favorite wino. A fireman put a pay envelope in the telephone booth across 4th Avenue from the fire hall. After awhile, along comes our favorite wino, bebopping down the street. He ducked into the telephone booth. As he started to use the phone, it rang and he picked up the receiver. The fireman on the line said, "Hey, I was just in that phone booth and lost my pay envelope! Look around there and see if you can find it."

We saw the wino lean over to pick up the envelope, then stuff it into his pocket. He got back on the phone and replied, "I don't see nothin'—ain't nothin' in here."

We watched him take the envelope out of his pocket, look into it and discover it was empty. Then he glanced across the street at the fire hall to see us all laughing.

The wino shook his fist clinched around the empty envelope and shouted, "You lousy firemen—I'll get you!" plus a lot more unprintable words. He shouted for about twenty minutes.

L. C. Tanley - Engine 29

A GREEN BUCKET AND IT'S YOURS

Another incident at the phone booth occurred when a fireman taped a dollar bill under the shelf. When a wino came by, the fireman called the number and the wino answered. The fireman identified himself as a disc jockey at one of the local radio stations. "We 've hidden a dollar bill in that phone booth, and if you can find it, you'll qualify for the next step to win five thousand dollars!" The wino was skeptical, but he looked around until he found the dollar and believed the caller. "I found it!" he hollered.

The fireman "disc jockey" said, "Okay, now you qualify." Seeing a green bucket on the sidewalk two blocks away, he added, "If you can find a green bucket and bring it back to the phone booth, you've won the five thousand."

The man looked all around. When he spotted that green bucket, he ran to get it. Then he discovered that the bucket was filled with concrete with a post stuck in it. It was used by parking lot attendants for the exit chain to hang on. The man struggled with the bucket, but there was no way he could lift it, so he tried to drag it back.

A parking lot attendant ran toward him, shouting, "That's my bucket! Where are you going with my bucket?" For ten minutes they stood there arguing over the bucket. Finally, the attendant gave up and let the wino take it.

He staggered up the street, dragging the bucket. By the time he'd reached the phone booth, the attendant accompanied by two policemen caught up with him. We watched the wino stand there, arguing with the men; at last they, too, gave up and left.

As soon as they were around the corner and out of sight, the fireman again rang the phone in the booth. "Okay, our man on the street tells us you have the bucket now. Here is a password—if you can find our man on the street and give him this password, he'll give you the five thousand. By the way, our man is dressed like a police officer. He's walking around there somewhere." The last we saw of that guy, he was running after those two cops. I wish I'd been there when he caught up with them and demanded his five thousand dollars!

Bob Barnard - Headquarters

A SLICK TRICK

One morning in early Spring of 1989, as we were preparing for shift change, a call come in for Rescue 1 and Medic 31 to respond to Shoney's on Dickerson Road. Immediately after they were dispatched, the operator announced for Engine 1 to respond also. Since the EMT on the engine was off duty that day I immediately informed the dispatcher that we did not have an EMT. He responded that we should go anyway.

When we arrived on the scene two minutes later, we found the restaurant surrounded by policemen. One of them asked me to follow him inside to observe the situation. We went to the kitchen. He pointed to the underside of the range hood and said, "There are his feet. How can we get him out?" A man stuck in the vent over the stove!

I decided to go to the roof to evaluate the situation from that angle. We found that this "would-be thief" was stuck in a one-foot-square vent. He had removed the cover and dropped down inside with his arms extended straight up over his head. His body completely filled the opening. I asked him if he could hold on to a hook pole for us to pull him up. He responded that he could hold to anything. We lowered a short hook pole to him and started pulling him out. When his shoulders were above the vent opening, a policeman standing close by said, "If you try to run after you get out I will shoot you."

"I'm not going anywhere," he replied, "I am worn out." He was completely covered with grease from head to toe. This made it easier to extricate him from this seemingly impossible situation.

Gordon Hall - Engine 1

SHE DID NOT FORGET

One winter night we were standing at the doors down in the "Bottoms." Somebody remarked, "It's so cold I hope we don't have anything tonight."

About that time a brand new white Oldsmobile slowed down. The door opened and a man pushed a pretty young girl out onto the sidewalk. She came limping into the firehall, wearing neither shoes nor coat.

Between sobs she told us her boyfriend had gotten mad at her and pushed her out. We told her to just sit down and wait, that he'd come back after her.

"Oh, no, he'll never come back. If he gets to drinking he'll never come back. He's got a case of whiskey in the trunk. I'll never see him again."

She was right. He did not come back.

At eleven o'clock we decided something had to be done. We questioned her about her home, and she told us she was from Alabama. We called the bus station for a schedule and the price of a ticket. All the firemen on duty chipped in. I gave her my jacket. With tears in her eyes she said, "I'll send the money back to you. I promise." We told her we were glad to help her and wished her good luck.

We had forgotten about it, but to our surprise three or four months later we received a letter with the money in it. There were many times we helped people in similar circumstances, but most of them we never heard from again.

Mack Johnson - Engine 14

THE MYSTERY OF THE FALSE ALARMS

The many false alarms that we had at the alarm box at 42nd and Charlotte were a mystery. They happened so many times that Assistant Chief Ike Moore came out from town to try to help solve the problem.

The alarm would be pulled two or three times straight one night, then it would be a week or two before it would happen again. We figured it was false every time but we had to check it out, nevertheless.

This was before street lights had been installed that far out Charlotte Pike and it was really dark out there. One night we arrived on the scene; it was a confirmed false alarm. One of the firemen just happened to shine a flashlight up a telephone pole that was right next to the fire alarm box. There was a telephone company work platform at the top of the pole. We found our culprit crouched down on the platform, watching all the confusion he had caused and enjoying every minute of it.

I guess you would say he was the original pole sitter.

D. White - Retiree

AN ACCIDENT THAT SHOULD NOT HAVE HAPPENED

The accident happened January 19, 1985. I was assigned to Engine Company 18 located on Gallatin Road near Douglas Avenue. Also assigned to this station was an ambulance, Medic 18. Often our engine would respond with them on calls. On this very cold dreary day we had six runs early in the shift. Three of these runs had been above ordinary fires. It had been pretty tough, but we went through it all right. In the wee hours the radio came on and gave the ambulance a call for an automobile wreck near Gallatin Road and Granada Avenue. I wondered why they had not sent our engine at the same time. Seconds later they did radio for us to respond also. Upon arrival we discovered there was a car wrapped around a pole and the lady inside was trapped and unconscious. My firefighters and the ambulance men worked closely together to tear open the car as fast as possible. We managed to get her out and into the ambulance. They were in the process of giving her life support procedures when someone said, "Cap, will you shut the door for us?" That's the last I remembered for a long time.

They told me later a little about what happened. A drunk driver in a truck plowed into the back of the ambulance, hitting me and a police officer. My firefighters thought that I had been killed instantly. I was rushed to Edgefield Hospital, a few blocks away; but on seeing the extent of my injuries, they immediately sent me to Vanderbilt Hospital where twenty doctors worked with me that night.

I have spent many long painful hours trying to recover. I can't walk or drive my car. My nerves had a lot of damage. I don't know what to do about it. I'll put it this way—I thank the Lord for being alive. That's why I stayed alive. The Lord helped me. I fully believe he had some reason. I don't know what the reason was but I believe it's there. I am fortunate to be alive.

Harry Williams - Retiree

BETTER TO LEAVE

One morning they salted Assistant Chief Willie P.'s coffee. He took a sip then turned to his driver and said, "Kimbro, does your coffee taste right?"

Kimbro said, "Yeah, Chief, mine tastes pretty good."

"Well I guess it's just me." He picked it up and drank the rest of it.

We all had to get up and get out of there to keep from laughing.

Headquarters Pow-Wow

FIREMEN CAN DO ANYTHING

Old Mrs. B. lived two doors from the firehall. She had a house full of junk—I mean junk! There were only little pathways where you could get through the house.

I don't remember whether this man was just visiting her or if he was a relative. Anyway, he had spent the night at Mrs. B.'s. One morning she rushed into the firehall.

"Would you all come up here and check this sleeping man in my back bedroom. He won't wake up. I've shook and shook him. He just won't wake up."

The Captain and the other firemen went to the back room to check him. They told her, "Why, Mrs. B., he isn't going to wake up! He's dead! He's frozen to death!"

She shrugged her shoulders, "He'll be alright. Just take him down to the firehall and thaw him out."

Billy Roy Carrigan - Engine 10

HE FOUND IT

It was about three o'clock in the morning when the alarm went out to Battalion 27, Battalion 21 and Engine 6 to respond to a house fire on Harding Place. Willard Fletcher was acting captain in charge of Engine 6, and the engineer was David Brown.

Willard reported flames visible. Not having spotted a fireplug nearby, he asked Firecom to have the next company in locate a plug and provide a water supply.

Then he jumped from the cab and immediately straddled a fireplug obscured by some bushes. As Gary Wilson advanced the 1¾ ″ line, he kept wondering why his acting Captain was slowly sliding to the ground with a grimace on his face.

So, for some time after that Willard was known as "Captain Fireplug."

John Petty

FROZEN IN

I've had some good times and I have had some rough times. There was a very stubborn fire in an old two-story house on West End. We would get it put out one place and it would break out again in another place. After an all-night battle in fifteen degrees below zero weather, we were exhausted, cold, wet and dirty. Our relief finally came at 6:00 a.m. When we got back to 17's I tried to take my coat off but the buckles were frozen. I heated a pan of water to pour on them to thaw them loose. My coat was frozen so stiff it stood alone on the floor.

At a time like that you tell yourself, ''It can't get much worse than this.''

D. White - Retiree

CALL THE CHIEF OFF

I was Chief McBride's driver. He brought his breakfast to the firehall every morning. I cooked it while he did the early morning paper work. The current joke at the hall was to put salt in somebody's coffee. I was sitting at the table next to the Chief, and Estell Flatt sat across the table. Just as he started to eat, Chief got a telephone call. I picked up the salt shaker and pretended to salt the Chief's coffee.

Estell said, ''You are chicken if you don't!''

With that I dumped about half a shaker into the coffee. As Chief returned Estell got tickled. I sat very straight faced. Chief took a big slurp of the coffee and spewed it all over the table.

He said, ''Estell, I'll get you for messing with my coffee!''

Three or four months later Estell came to me. ''Please go tell Chief McBride you salted his coffee. He's been on my case long enough.''

Jerry West - Training Center

AN UNSUSPECTING *GOOD OLD MAN*

Bobby and I used to pull some good ones on Mr. Bill Nemon. He was a good old man. He even paid for my lunch the first two or three months I was on the Fire Department, but we still couldn't resist playing pranks on him.

Remember when milk cartons first came out, they looked like they had candle wax on the outside of them. Mr. Nemon would bring milk and put it in the refrigerator. Bobby and I took my pinpoint knife and punched a tiny hole in the corner of the carton. We'd drain his milk out. Then we would take a little warm candle wax and close up the hole.

Mr. Nemon couldn't figure out what happened and he never did. He died thinking something very strange was going on.

George Owen - Headquarters

WHAT A NIGHT

I was working on Engine 10. One night when we got to work the operator told us to go to the funeral home, and then back up everybody else so they could go. When we returned we were told that 13's had a fire. We went to back them up. When they returned to the hall they had used 15 or 16 sections of 1¼ hose. While we loaded their engine they went in and cleaned up to go to the funeral home. After they returned, the operator sent us to 4's to back them up. They had a lay out. They used every section of hose they had on their engine. We loaded it back on the engine for them. About that time 18's went out. Guess what? The operator sent us to back them up. When they got back, we loaded their engine.

When we finally got back to Headquarters about midnight, I said, "Well, all we need now is to have to lay out our own hose."

Captain Hobby Hall wearily warned, "Don't say that."

About two o'clock they rang the bell. They hollered, "Engine 10". When we turned on Gay Street it looked like the sun was coming up in West Nashville. Two houses were on fire! We dropped about 15 sections of 2½ inch hose. We got in about 5 a.m. We had to load up that 15 sections.

We loaded 4 or 5 engines in one 12-hour shift. They wore us out!

Jerry West - Training Center

SOME FRIEND

Most of the new guys that would travel down at the "Bottoms" would wind up staying soaking wet all day because the firemen threw water on them. I traveled down there one day. I was sitting there and all of a sudden about five gallons of water hit me from nowhere. Then one of the guys came up and he was going to be nice. He told me he didn't know why they did such as that. He acted like he wanted to be my friend. He said, "I'll take your shirt and hang it up and let it dry." He took it and hung it in the cooler. When he brought it back to me it was frozen.

Somehow I had a suspicion that this "friend" was also the one who threw the five gallons of water in the first place.

Wayne Vick - Engine 1

HELP ME SOMEBODY

We had a real nice fellow for an engineer at 22's. He used to brag, "Nobody is going to take a plug away from me after I get to the scene. No sir, nobody's going to get my plug and they sure are not going to take my engine away from me."

We kept egging him on, "Oh, yeah, they will if they want to bad enough."

One night it happened. We had a call to the third floor of Middle Tennessee Mental Health Hospital. We went there to check it out and left the engineer on the engine. When we came down the steps we saw him running around the fire engine hollering, "Hey somebody! Hey somebody!"

We rushed over to find a lady without a rag of clothes on, sitting up in the driver's seat.

We teased him, "Don't feel so bad; at least it took a naked lady to take your engine."

George Owen - Retiree

NO WATER—NO LINE—NOTHING

One night down on Charlotte a ladder fell with me. I was up on the ladder with a booster line in my hand. I could hear the engine going "Burrrr." That meant no water was coming out. I was trying to hold on to the line, get the window up and balance myself on the ladder.

About the time I popped the window up, the ladder decided to slide off the wall. I quickly decided it was a lot easier to jump in the window than to ride that ladder two stories to the ground. I jumped through the window. Just as I landed the fire decided to come into that room. There I stood with just my helmet—no line, no water, nothing!

I could hear Captain Hobby Hall and the other firemen on our engine hollering. They were trying to get the ladder back to me. It had fallen down and gotten tangled up in a clothesline.

I heard Sam Kittrell off of Engine 7 coming up the stairs. I yelled for them to come to me.

Sam said, "What in the hell are you doing up here without a line?"

I said, "I had one till I came through the damn window!"

Jerry West - Training Center

SNAKES ALIVE!

One cold December day, in the dead of winter, Captain Charles M. Vance answered the main line at Engine 6. He was told to respond non-emergency to a residence about three blocks from the hall. Charlie asked the dispatcher the nature of the call. He was told some little old lady was hearing hissing noises coming from the gas-fired logs in her fireplace, but she could not smell any leaking gas.

Charlie sent firefighters John Petty and Willard Fletcher from the Chemical Unit to investigate. The two men arrived, introduced themselves, and asked to be shown to the fireplace. Sure enough they, too, heard a hissing sound but could not locate the source. Then as the two men were crawling on the floor the little lady remarked, "You know my son lost his boa constrictor last week. I'll bet that thing crawled back in there."

Just about that time the snake stuck its head out from a crevice behind the mantel. John and Willard nearly wrecked the living room in their haste to depart. The son was summoned to quickly capture the reptile.

John Petty

I JUST CAN'T DO IT

It took two or three weeks to work out all the details of this prank.

We had a fireman who had a rare type of blood but he went berserk at the thought of donating one drop of it.

We were all in the back listening when the prank call came. An hysterical woman called for him. She said, "Is there any way in the world I can get you to give my husband two pints of blood? The doctor said he will die without a transfusion. It's so hard to find anyone who has this rare type of blood."

The fireman said seriously, "Ma'am, I'd like to help you, but I just can't give blood. It makes me deathly sick. I'll tell you what I'll do. I'll check around and see if I can find somebody with that type of blood. You give me your telephone number. I'll call you back."

"It's hard to catch me, I'm at the hospital so much. I'll just call you in a few days," she said.

Then we got a fast-talking fireman to imitate a lawyer. He call the fireman at home on his day off. Knowing how he loved money, the "lawyer" told him, "I'm prepared to pay you $500.00 if you will just give this man a pint of blood now and another pint in six weeks. I'll give you some time to think about it. I'll call you back."

The next day when the fireman came to work, he told us about the lawyer calling. As he paced the floor all day, he was pale as a ghost. His stomach was upset and he couldn't eat his meals.

We realized that this had gone far enough, so we told him it was a joke.

He said, "I sure am glad. I don't think I could have gone through with it, even for $500.00."

Georgc Owen - Headquarters

For many years firemen spearheaded the "Toys for Tots" campaign, collecting, repairing and delivering toys for needy children.

JINGLE! JINGLE!

For many years the Nashville Fire Department took great pride in their "Toys for Tots" program. We repaired and distributed toys for underprivileged children at Christmas time.

On my day off I had to go to a funeral home near 17's where I worked. I decided to go by the hall to see how the toy project was coming along.

When I got there the engine was out on a run so I just went on down to the basement where we worked on the toys. As I went down the steps I stumbled over a tiny rusty bell. It had a wire attached to it. The thought crossed my mind that sometime it might come in handy for a prank.

That time and situation presented itself immediately, as I came back up the basement steps. The way we parked our cars, made the hoods hang slightly over the basement steps. There I stood looking right up under Hilderbrand's car. I thought how funny it would sound with Hilderbrand driving down the road with that little rusty bell ringing. I reached up and attached it to one of the wires under his car. It was easy!

A few days later Hilderbrand came in complaining about a funny sound in the motor of his car. We all sympathized with him.

He told me later that he traded that car off because he couldn't locate that funny sound.

D. White - Retiree

GO AHEAD

We played the old "billfold on the sidewalk, attached to a fishing line" trick many times. One time it went a little further than we'd planned.

About dark one night a "real know-it-all" kind of a guy dressed in a three-piece suit, came prancing down the sidewalk in front of the firehall. He pretended not to see the billfold until he got right even with it. All of a sudden he slapped his big number eleven down on it. We jerked the fishing line, but it was too late—the line broke.

The guy picked up the billfold and started looking through it when a fireman on the PA system blared out, "Put that down!" It scared him so bad he dropped it and started looking around to see where that voice came from.

"Turn around! Put your hands up! This could be the FBI!" It sounded official. Several firemen went out and told him they were checking to see if he was armed.

He looked so funny, standing there on 4th Avenue with his hands in the air, that we all got tickled. Somebody said, "Now do the hokey-pokey."

This made him so mad that he threatened to go tell the mayor.

The firemen said, "Oh, go right ahead—we got him last week!"

Bob Barnard - Headquarters

WE GOTTA EAT

A fireman got laid off one day. He went home, piled his wife and kids in his car, and drove over to the Chief's house. They were sitting on the porch when the Chief got home from work.

"What are you doing here?" he asked.

"We came for supper. You laid me off. My family's got to eat somewhere."

The Chief put him back to work that night and even tried to loan him some money.

George Owen - Headquarters

AN UNWELCOME GUEST

This happened at Old 1's. This firehall is located in a unique neighborhood. It was once a high class residential area but it is now predominantly commercial. We did have good rapport with the few neighbors we had left—so much so that it was our practice to leave the doors open when we went out on a run.

We had been out on a fire practically all day. It was suppertime when we came in. We were all cold, wet, dirty, and hungry. After we got the engine cleaned up and got cleaned up ourselves, we gathered back in the kitchen to eat. About bedtime I went up front to the office to do paperwork. All of a sudden I heard a loud scream. "Little Boogie" Friedman came running up through the firehall hollering, "Cap! Come here quick. There is somebody in my bed!"

I went back to the bedroom to find Boogie's blanket and spread in the floor. The room reeked with the nauseating smell of whiskey and urine. There was a big man in the middle of Boogie's bed.

I shook him. "Hey! Wake up! What are you doing in here?"

He raised up about halfway with a silly grin on his face. "What's the matter? Did I get in the wrong bed?"

"You sure did. Get up and get out of here."

He was fumbling around putting his shirt on and stumbling toward the door.

I said, "Wait a minute, you've got to clean up over there where you've wet on the floor."

He took off his snirt and wiped up the floor, put his shirt back on and took off into the night.

This incident cost the Fire Department the price of a new mattress. Little Boogie refused to sleep on his old mattress ever again.

Gordon Hall - Engine 1

MY FIRST THREE DAYS

It all started in late 1946. I was recently discharged from the old Army Air Corps of World War II, out of a job, with a wife and one child and expecting another in July. The thought of being a fireman (firefighter by today's standards) did not appeal to me at all.

My stepfather, B. D. Skinnier, had been a fireman for five years.

One evening, while eating supper, my mother asked me if I had thought about being a fireman. I told her that I wanted to be a pilot, since I was already a private pilot working on my commercial rating. Being a loving parent and wise beyond my years, she suggested that I put in my application with the understanding that if I was hired and did not like the work, or if I found something better, I could always quit. Needless to say I took her advice and was hired July 7, 1947. To the best of my knowledge the city hired ten recruits they called "floaters."

We were to work any shift, anytime needed. As a result, my first paycheck was more than Assistant Chief Finley's (because of extra hours and double shifts) although my regular pay was $180.00 a month, 84 hours a week with one day off every third week.

On day one, I reported to Engine Company 5, Captain Abe McKinley at fire headquarters, 5th and Deadrick Street. Captain Abe instructed fireman Buck Primm to get me turn-out clothes and show me where to ride. I was instructed, "When we get to the fire you stay with me and do what I tell you," by Captain McKinley. Well, day one was uneventful.

Day two, I reported to Lieutenant L. C. Young at Engine Company 9 at 4th and McGavock Street. After introductions all around, Les asked me if I could slide the pole. I told him I could. He replied, "This I gotta see."

I slid the pole and I burned the devil out of my hands, but I wouldn't let on that it hurt. I think he knew, but he just told me how to improve my next slide. Still no fire calls.

Day three, I reported to Captain Lou Dows of Engine Company 15 at 12th and Caldwell Avenue. I was filling in for my stepfather, who had the day off. My stepfather and I were about the same size, so I was using his turn-out gear.

About one o'clock the ticker tape began clicking off *Box 551-Purina Mills- 4th and Chestnut.* Engine 15's territory. Away we went. Standing on the tail board behind the driver and looking through the windshield, I watched traffic give way as we approached with siren blaring and Captain Dows ringing the bell mounted on the fender by him. The adrenalin was pumping fast as we turned the last corner at 4th and Chestnut and approached the scene where other companies with flashing lights were arriving.

We were lucky. No fire, just a dust alarm. Returning to the station little kids would run to the curb and wave. We were their heroes and from then on I was addicted. Something got in my blood and it lasted for 38 years.

Jim Cundiff - Retiree

DEADLY GAS

This happened in North Nashville. We received a call that said two kids had gone down into an abandoned cistern.

When we arrived on the scene Chief Linger squeezed his way down into the cistern. He immediately passed out.

Another fireman, I don't remember his name, went down. He passed out also. Now we had two children, Chief Linger and another fireman down there, all thought to be dead.

The rest of us got all the men standing around to help us lift the big concrete slab off the top of the cistern.

Captain "Pea Nose" Loring went down and brought Chief Linger out first. He was pronounced dead. After we worked with him for a while he was revived. They revived the other fireman also. The kids were both dead—methane gas.

Billy Roy Carrigan - Engine 10

A RETIRED FIREMAN REFLECTS THANKFULLY

Twenty-five years ago I sat across the desk from Mayor Ben West; I had just been hired as a recruit for the Fire Department. When asked, "Why do you want on the Fire Department?" I answered, "To serve the citizens of Nashville in return for job security and a good pension plan." It seems like such a short time ago.

Now as an "old-time veteran" I can look back over the years with a feeling of satisfaction. I feel that I have served well and am now ready for the first pension check.

I know each time I pass a fire hall or see a fire engine racing down the street, many memories will come to my mind, such as my first emergency run to the box at Humphreys and Chestnut Streets, the first charred body I helped carry out of a burning building, fires like the Exchange Building, the State Fair grandstand, and the Maxwell House.

I will always remember the little bright-eyed boys as they surveyed our equipment, and their saying, "When I grow up I want to be a fireman."

Thank you, Nashville, for giving me the opportunity to serve you.

I thank you for helping my wife and me raise good citizens. As I settle back now as "Mr. Private Citizen," I, too, can relax knowing I am protected by the greatest Fire Department of them all.

Joe Huber - Retiree

NO PONDS FOR US

During the 40s and 50s era, a lot of homeowners installed ponds on their property. Some stocked them with goldfish, some with water lilies, etc. My wife was no exception. She wanted a pond in our backyard. After a few days of discussion and planning, we began excavation.

Meanwhile, John Owens and I were working our fire extinguisher route on West End Avenue. We were stopped at 28th and West End headed downtown, when we heard the siren. We waited and watched as the red light approached and the siren grew louder. Car 104 with Battalion Chief Jesse Vaughn and his driver flashed by. John said, "Let's see where they go." He made a sweeping U-turn in West End traffic and fell in behind 104.

We pulled up on the scene as Captain Jesse and his driver emerged from their car. Captain spotted John and me and told us to follow him. Back then no EMTs were available, so Engine Companies did not respond to medical emergencies. When we arrived in the backyard, I was horrified to see a small 2½-year-old baby lying face down in a fish pond with about six inches of water in it.

I immediately pulled her out and started artificial respiration. We did not know CPR as we do today. I was astride her back with her lying face down on the ground. It seemed like an eternity before the ambulance arrived, although I was told later it was a matter of three or four minutes.

The attendants picked me and the baby up and placed us in the ambulance. All the way to Saint Thomas Hospital, I never quit the artificial respiration. John was in the ambulance with me and he kept me from falling, as we zig-zagged through traffic. At one time John told me he thought she was responding but, unfortunately, the doctors pronounced her D.O.A. I tried to console the grieving mother but the doctors had to sedate her. It was heartbreaking!

That evening at the Cundiff home the pond in the backyard was filled in. This tragedy would not be repeated with my 2½-year-old daughter, nor my 7-month-old son. My wife would be spared the experience of a possible similar incident. It's an experience one never forgets and it haunts me to this day. When you ask yourself *Did I do everything I could?* the answer has to be *It was God's will.*

Jim Cundiff - Retiree

YOUR DECISION

You work in a single station fire company. Early in the morning hours the bell rings. You raise straight up in bed, turn and your feet slip into your boots by your bed. You stand and, pulling your bunker pants up, you race out into the apparatus room. As you step on the engine it starts and pulls out at full speed. The cold January wind hits you in the face as you struggle to put your coat on. Off in the distance you can see a red glow in the sky. You know you have a working fire. Off in the distance you hear other sirens and you thank God you live in a town that has more than one fire engine. As you turn the corner yellow pine smoke fills the street. Your eyes see the figure of a woman in her robe wringing her hands and crying, "My baby is up there." You look up and heavy smoke is filling the window. You know in your heart nothing could be alive in the house. Just then the ladder company places a ladder in the window and up you climb. When you reach the top you cannot see anything—not even if there is still a floor in the room. You look down at the young mother still pleading to get her baby. In the window you go. Crawling on the floor, you pray you don't find what you are looking for. Just as you think your lungs will burst, your hand finds a limp body. Holding the child in your arms, you start for the dim light you know is the window. Just as you hand the child out to another fireman on the ladder you look down and see the mother crying, "Why-why?" Right then you know there must be a better way to save lives than after the fire starts. You make up your mind to get into fire prevention.

John Collins - Retired Fire Marshal

YOU CAN GET PAID TWICE

Harry brought two lawn mowers to 23's for me to repair. He told me if I'd fix one for him I could have the other one. I picked out the best one for myself and I fixed the junky one for him. After I got it to running pretty good I told "Bumble Bee" to put a can of gas behind a bush.

I said, "Harry, let's go out there and see how you like the way the lawn mower works."

After making two or three laps around the lot, he said, "Boy, it cuts good!"

I said, "Well, we have got to cut the grass and besides that you can get some exercise."

I went into the fire hall and called down to 13's and 17's and told them to call Harry several times about some coins. Harry was a coin dealer. The calls began to come in. When he would come in to answer the phone we would fill the gas tank up. He cut that whole big lot with just one tank of gas.

He said, "Boy, that sure gets good gas mileage!" Before he left work I screwed the choke screw all the way in. That sucker was burning more gas than it ever did before.

He went home and cut his little yard that was about the size of a good sized living room. He ran out of gas.

He brought it back. He said, "I can't figure it out. I cut this whole big lot and now I can't cut my little yard." I said, "Well, Harry, you jarred it up on the way home."

He said, "Jarred it up! I don't understand that."

I agreed to fix it again for two silver dollars, which I later sold back to him for $10.00 apiece.

George Owen - Headquarters

SNOW AND MORE SNOW

We were working at Old Headquarters. It snowed so hard we could hardly see. Captain Young and Captain Escue had us shovel snow off the ramp. We shoveled until about 3 a.m. Finally they told us we could knock off and leave the rest for the morning shift.

We decided we'd get even with them for making us stay up all night. We waited until Young and Escue went to bed. We went back outside and shoveled all that snow back onto the ramp.

Captain Young was called into the Chief's office to explain how the snow got back on the ramp.

Nobody knew!

George Owen - Headquarters

WHERE ARE THE NUTS?

I was working at Old Headquarters on Deadrick Street. Chief Ragsdale came in carrying a 30 or 40 pound sack of pecans. He said, "Now don't you men get into these nuts. I've got to give them away." He hung them up on a hook.

Well, we couldn't stand that. We got a nail and worked the mesh on the bottom of the sack so there was a hole big enough for us to get a pecan out. We emptied the sack that night. Then we took the nail and worked the mesh back into place.

When Chief Ragsdale came in the next morning and found all the nuts gone, he was upset! He took the empty sack into his office.

"I can't figure this out. The sack has not been untied and there isn't a hole in the sack. How could this sack be empty?" he asked the secretary. She didn't have an answer.

He didn't ask us and we didn't tell him.

Bill Stoutt - Retiree

SAD MEMORIES

One call that brings back sad memories was the night that Captain Harry Williams was hurt.

Engine 18 responded to an auto accident on Gallatin Road about one fourth mile from the firehall. A lady had hit a telephone pole. The firemen had removed the lady from the car, and in the back of the ambulance paramedics were doing life-saving procedures on her. Captain

Williams and a policeman were standing at the back of the ambulance looking in, when a driver with a long history of DUI ran into them. Severe injury to both men and damage to the ambulance was the result.

I was the second unit on the scene. They called us to transport the police officer to General Hospital. He had two broken femurs and other injuries, but he had remained conscious the whole time. We splinted his legs and transported him to the hospital where they listed him in stable but serious condition.

As soon as we dropped this patient off we got a call to go to Edgefield Emergency Room to pick up Captain Williams and take him to Vanderbilt Hospital. They had him stabilized but he was delirious and had to be restrained. All the way to Vanderbilt I tried to get him to remember my name. It was no use. He couldn't remember his own name. He did have trauma to the head, with a big laceration and several fractures.

At Vanderbilt, they took him straight to surgery. After surgery he was to stay in Intensive Care for more than three months.

What got me most about Captain Harry was when I checked his pupils. One pupil would not move, which was a firm indication of brain damage. When the DUI had crashed into him, Harry's head hit the back window, causing blood and fluid to swell his brain. That would explain why the pupils were under-active. I always worried about that.

Only five hours before the accident, I had been talking and joking with Captain Williams. This accident could have happened to anyone of us. It made me realize how serious our job really is. It made me think: "Would I be able to come home the next morning? When I put on this uniform, will I be able to take it off the next morning."

Henry Booker - Rescue 1

SCARED OF A KITTEN

Mrs. Cato was Chief Ragsdale's secretary. She was a very nice lady. We could not resist bugging her. One day a scrawny kitten came wandering in the firehall. I put it in Mrs. Cato's big side drawer. Later that day she reached in that drawer, without looking, to get something. She touched that soft warm fur. She started screaming and crying and ran into the Chief's office.

Chief Ragsdale came out into the hall. He said, "I want you all to quit bugging her. I can't get a bit of work out of her."

I would like to say we stopped; but she was right there and a perfect target.

Bill Stoutt - Retiree

FIRE PREVENTION

Chief Dan Hicks, was known as the father of Fire Prevention in Nashville. At that time Fire Prevention activities increased and Nashville was made aware of good fire prevention practices. Fire Prevention week meant poster and slogan contests, firemen visiting the schools and Nashville's biggest parade. A highlight was the crowning of the "King and Queen of Nashville's Fire Prevention Week." One year Pat Boone was crowned king. Later came the "Little Miss Spark" and "Miss Fire Prevention" contests. One of the last to hold the title of "Miss Fire Prevention" for Nashville was Ms. Oprah Winfrey.

When Metro Nashville was chartered, the Fire Marshal's office was enlarged and located in a back room at Fire Headquarters under the jail on 2nd Avenue No.

Paul Roberts was the next Fire Marshal, and the arson squad was formed to help reduce fires by investigating all structure fires in Nashville.

During these years Nashville was judged in Fire Prevention by submitting scrap books to the National Fire Protection Association in Boston, Massachusetts.

Then Fire Prevention *week* grew from one week in October to a year-round program. "Fix-Up-Clean-Up-Paint-Up" in the Spring, programs for civic clubs, Scout troops, PTA, church groups, films, talks, demonstrations and Toys for Tots—you name it, the Fire Marshal's office did it.

The next Fire Marshal was Howard Boyd—then John Collins and William Perry.

The Fire Marshal's office was relocated to the old home of Engine 4 and Ladder 3 at 5th and Russell Streets. It is known as the Fire Prevention Bureau. This building is now named for and dedicated to Mary

Courtesy of Metro Nashville Archives

Miss Oprah Winfrey receiving the *Miss Fire Prevention* crown in 1971.

Catherine Strobel, the first female hired by the Nashville Fire Department as a clerk in the office of the Fire Marshal.

The dedication, concern and desire to have a fire-safe community continues. While we must have a ready, front line of defense against fire—the suppression forces—we, like the medical profession, understand that it is far better to prevent than react after the fact.

"Saving lives is prevention. Saving property is suppression."

John Collins - Retired Fire Marshal

MEMORIES

November 1, 1951 was a memorable day for my wife, our five children, and me. That was the day Mayor Ben West gave me a job with the Fire Department. It was the start of twenty-five years of making a decent living in a pleasant atmosphere. During these years I experienced many events—some pleasant and some not so pleasant.

My first run was a box alarm at Humphry and Chestnut Streets. It was just a grass fire but that made me a seasoned veteran. That was the first of 5,570 runs of varying descriptions.

My first multi-storied building fire was the old Exchange Building on Church Street between Third and Fourth Avenues.

The hottest fires I experienced were at the Fairgrounds. One was a large mule barn and the other was the grandstand.

The Maxwell House fire made national news because of the history of the building.

During my last year on the fire department, I worked fires that resulted in eight deaths. This is something that one never gets used to. Some of the "runs" turned out to be downright funny or pathetic, or even heartwarming. I shall never forget the day my company was ordered to the campus at David Lipscomb to replace a flagpole rope. While we were doing so, a teacher brought her fourth grade class out to watch us. The teacher introduced me to a little Oriental girl and said, "Suzie wanted to be a nurse yesterday, but today she wants to be a fireman."

One night we made a fire call at 12th Avenue South and Bate Avenue. The fire was on the second floor. A drunken tenant was determined we were not going to put the fire out. I guess all of our axes and hoses scared him. As soon as we got water I literally washed him down the steps. After we extinguished the fire, I told him that our rule was to put the fire out at all costs. His attitude then changed and he cried out, "Bless you, you saved the lives of my family!"

Once I was mistreated by a senior officer, but I "got even" with him. When he was later hospitalized twice I gave him blood each time.

A preacher-fireman friend of mine once said of himself, "On my job with the Fire Department I am a saver of lives, but in the pulpit I am a saver of souls."

The pay wasn't the best in the world but I would not have swapped jobs with anyone.

Job Huber - Retiree

HARVEY'S HAD IT

In the last week of June 1946 Nashvillians were horrified by the news that the "pace-setting" Harvey's Department Store on Church Street had experienced a destructive fire.

A few days later a full page in both local newspapers carried a letter from Mr. Fred Harvey, addressed to "All the Friends of Harvey's All Over the Southland." The letter appeared in the paper on July 7, 1946, a Sunday morning. A reproduction of part of the letter follows:

My Friends:

To my own employees, to city officials, to the radio stations, the newspapers, and to hundreds of citizens who were of invaluable assistance during and after the fire, I am forever indebted. A very special vote of thanks must go to Fire Chiefs Demonbreun and Ragsdale and the courageous men of the Nashville Fire Department. It was due entirely to their efforts and ingenuity that the fire was confined to one floor of our West Building and that the final losses sustained are much smaller than was at first feared. Under difficult circumstances and with inadequate equipment they did a job nothing short of phenomenal. Nashville can well be proud of such efficient and courageous public servants.

Faithfully Yours,
FRED HARVEY

A fair appraisal of the years intervening between the date of the above letter and the time of this writing will amply sustain the premise that Mr. Harvey, Sr., and Mr. Harvey, Jr., have both made substantial contributions to the economic welfare and character of this entire region.

High rise fires, even in small buildings, are very dangerous and difficult to combat. The taller buildings of today have built-in fire defenses in the form of standpipes and sprinklers with water pressures being beefed up by auxiliary pumps. The Harvey fire served for many years, and still does, as an example of the problems posed by this type of emergency and the attack methods to subdue and control it.

Retired Chief William C. McIntryre

YOU CAN'T WORK HERE

I started out kinda peculiar like. My first assignment was Engine 5. I went down there to report to work. My father-in-law, "Poopey" Artie

Burris, was captain. When I went in the door he looked up at me and said, "What in blazes are you doing down here?"

"They sent me down here to work."

He shook his head in disbelief. "I can't take that. You got no business being here. We are kin people. You got no business at this hall."

He went up to fire headquarters and told Chief Demonbruen, "You got to do something with him. He's kin to me. He's got no business working at my station."

Chief Demonbruen wrote out another assignment to Engine 6. That's where I started. "Old Poopey, he's a good fellow."

Harry Williams - Retiree

WATCH MY COFFEE

Homer always fixed his breakfast at Old Headquarters. He was just starting to eat when they came in and told him he had to move his car.

He told the Chief, "You watch my coffee."

When he came back he took a big drink of coffee. It was so salty he choked and sputtered, "Chief, I told you to watch my coffee!"

"I did. I watched them put the salt in it and I watched them stir it up."

Jerry West - Training Center

Courtesy of Mike Dubose

Bill Radcliff, Engine 3, and friend. (A modern day "Sparky" in inset.).

3 Animals

SPARKY

During the years 1963 to 1972 I was assigned to the Fire Marshal's office. One morning during Fire Prevention Week I was working in the Hillsboro area and I had ten minutes to reach a preschool class in North Nashville. I was not told the assignment, but I was the only one who answered the radio, so I was elected.

When I arrived at my destination I was dumbfounded. I was asked by the teachers to give a presentation on Fire Prevention. I was terrified; I had never made a speech in my life. Well, maybe as a kid when I had time to prepare my lessons. I said a silent prayer and asked the Lord to please see me through this ordeal and not let me fall flat on my face.

With that, I backed up to the stairs, sat down on the steps and surveyed my audience of about 30 boys and girls, 5 and 6 years old, all seated on the floor. The front door looked very inviting, and believe me I wanted to run out. A four-alarm fire would have been less terrifying at that time. But I stood my ground.

"Boys and girls," I said, "how many of you have heard of Sparky the Fire Dog?" Hands went up all over the room. "Now what?" I said to myself. The idea hit!

"How many of you know how Sparky received his name?" No one? Well let's see.

It was a dark and stormy night, the lightning flashed and the wind was howling like a banshee as the little spotted pup struggled along the deep rutted road. His head was nearly dragging the ground, he was cold and hungry and he had no place to stay. He was saddened by the fact that his mother had been killed in an accident earlier, making him a orphan.

Suddenly, the little dog raised his head as he heard the clamor of hooves and a bell ringing in his ears. Around the corner came the horses, three abreast, pulling a wagon belching smoke and steam as the fire engine sped by. He paused for only a moment to take in the scene as the firemen on the back of the engine held on, with their coattails flying in the wind. The puppy did not tarry as another bolt of lightning and a loud clap of thunder reminded him that he needed to find shelter.

Turning the corner he spotted the open door of the fire station and a roaring fire in the fireplace. *Oh! Boy!* he thought, a place to get warm and dry. Cautiously he stuck his head inside. Nothing happened. He moved a little closer until he could feel the warmth of the fire, and he lay down. He was fast asleep when a sudden gust of wind swooshed down the chimney, scattering sparks and live coals across the wood floor. The little dog, still damp from the rain, playfully patted at the sparks with his wet paws. Hiss, and the spark was out and so on to the next one.

Later the firemen returned to the station and found the puppy lying by the fire.

"Well, what have we here?" said one of the firemen. "He's a cute little fellow, Captain, can we keep him?" asked another.

"Well, I guess we could use a mascot," Captain Bill responded. "Looks like he named himself. See all the charred spots on the floor. Looks like he saved our fire station by putting out all those sparks. So, why not call him *Sparky*. The vote was unanimous. The time was October 9, 1871. The firemen had returned from fighting the Chicago fire. Some years later Congress would enact a law making the week of October 9th "Fire Prevention Week" in remembrance of the Chicago fire, and *Sparky* was designated the official fire department symbol and mascot.

Today Sparky is still helping firefighters all over the world combat man's greatest enemy, untamed fire.

Jim Cundiff - Retiree

SPOT

In 1913 Chief Rozetta was presented a mascot dog by Victor Coode. The puppy was a black and white spotted English coach dog, appropriately named "Spot".

Spot grew up having an unbelievable fondness for the horses of Hook and Ladder 2. Every morning when the horses were brushed and curried, he patiently waited in line for his grooming.

Hook and Ladder 2 never left the hall without him. After they reached a fire, he followed right with the firemen. He did not climb the ladder but would stand at the bottom and paw the lower rung. His ability to stand heat and smoke was remarkable.

A firehorse once stepped on his leg and broke it. This did not stop Spot—he still went to the fires, with his leg in a cast.

ECHO

Only the fittest and most capable horses in the Fire Department were chosen to drive the Fire Chief. Echo had the honor of being the favorite of two Fire Chiefs.

Echo's beginnings as a firehorse are shrouded in mystery—no fire department records state where, when or even how he became the horse that pulled Chief E. M (''Mon,'' for Monroe) Carell's buggy.

Perhaps the big chestnut bay was a gift to Chief Carell. Chief Carell's fearlessness had won the hearts of Nashvillians, although he was a modest man—''a plain, blunt, generous-hearted official.'' To the firemen he had earned the reputation of a leader who did not send his men ''into a fire'' but rather led the way, saying to them ''Come on!'' It was this character trait that cost him his life.

At 3:00 a.m. on a frosty February 28, 1900, Echo, driven by Charles Benson, thundered through the darkness on the slippery cobblestone streets to Chief Carell's home on Sumner Street (now Fifth and Cedar Streets). After arousing the Chief, they headed across the bridge to answer a fire call at 504 North First Street.

Benson was not driving fast enough to suit the Chief. Forgetting his personal safety in his determination to get to the fire, he took the reins near College Street (now Third Avenue). Crossing the square, with sparks flying from his steel-shod hooves, Echo was going at tremendous speed. The buggy was swaying behind like a paper box. All of a sudden the buggy struck a post which supported trolley car wires at the Market Street corner of the Courthouse. The post caught the buggy between the wheel and the shaft. Shaves and harness broke. Free and unhurt, Echo trotted back to the fire hall. Chief Carell and Benson were thrown headlong twenty or thirty feet over the dashboard. They both lay unconscious in the street. Benson, although badly bruised, was not seriously hurt. Heroic Chief Carell died seven days later from head injuries received in the wreck.

Chief Carell lay in state at Fire Hall 2. At the funeral, everyone wondered how such an unlikely accident could happen to a veteran Chief and an experienced firehorse.

Spared the awful task of pulling his master's casket from the fire hall to the church, Echo was not mentioned at the funeral. He was never penalized but continued to be loved by Nashvillians as well as all the firemen.

Echo continued to serve in the Fire Department for many years.

A. A. (Tony) Rozetta succeeded Chief Carell and Echo was his favorite horse. Years later Chief Rozetta said, "That horse had sense. That was one horse that was thought so much of he was pensioned. He was given rich pasture in summer and shelter in winter."

The date of Echo's death was not recorded but the circumstances brought tears to the eyes of his Chief and sorrow to many Nashvillians. It was a crisp, frosty fall day and Echo, now very feeble, was being led along familiar streets from one fire hall to another. Suddenly he just dropped down on the street and the firemen thought he was dead. Word was sent at once to Chief Rozetta, who came as fast as possible to his pet. At the sound of his Chief's voice, the noble steed raised his head and looked lovingly into his master's eyes. He made one great effort to rise before falling back dead.

FIREHALL DAN

Firehall Dan was coal black "part Arabian and part Percheon, but not broad-backed." He was horse number twenty- six on the Nashville Fire Department records. In 1916 he was fifteen years old. He and his partner, Dash, worked on Engine Team 9.

They were separated when Dan was sold to George E. Harris, who operated a grocery store located at Acklend and Twenty-first Avenues. Instead of pulling the bright shiny red fire engine at breakneck speed through darkness, sun, rain, or snow to a fire, Dan's job now was to pull a market wagon to deliver groceries.

Instead of the brave firemen, intent on getting to a fire as quickly as possible, his driver now was a thirteen-year-old boy whose job it was to deliver groceries to homes in South Nashville. Dan plodded wearily along, stopping at the curb in front of the houses and waiting patiently while the boy carried the baskets of groceries into the kitchens. Dan stood and waited, but he remembered the former days when he and Dash thundered along side by side with their heads high and their tails and manes flying in the wind.

On August 18, 1916, Dan and his young driver left the grocery with a wagon loaded with sixty slat baskets full of groceries. About eleven o'clock they reached Sixteenth Avenue South. Dan stood waiting patiently while the boy went around to the back of the house.

Dan heard the wail of the new-fangled engine from Seventeenth and Broad coming closer and closer. It stirred his memory. As the engine passed him, Dan jerked into action. Twelve to fifteen baskets of groceries scattered into the street but no matter to Dan, he was in his glory again. His heart was pumping and he was breathing hard but he was catching up with the engine. He was so close the firemen could have reached out and touched him.

A tragic thing happened as the engine and Dan turned the corner at old Ward-Belmont College. The wheel of the wagon caught in the street car track. The wagon turned over and down went Dan. The other forty-five baskets of groceries spilled out onto the street.

Not to be outdone, with a mighty heave Dan struggled to his feet. Although the staves of the wagon were broken and it was otherwise badly damaged, the wagon righted itself. Dan raced with the now empty wagon to catch up with the engine.

When Dan's driver reached the fire he found the fire had been put out. Dan and the firemen were having a grand reunion.

Two years after his "run away," Mr. Harris located Dan's old partner, Dash. He was owned by Mr. D. D. Canfield who operated the Model Laundry. Mr. Harris sold Dan to Mr. Canfield. Dan and Dash were reunited. Their old age was spent happily pulling the laundry wagons together through the familiar Nashville streets.

POOCH

A very special lady firefighter resided at Fire Hall 13, long before lady firefighters were in vogue. She endeared herself to all the firemen.

Residents of West Nashville grew accustomed to looking for the little brown and white terrier, "Pooch," standing on the platform, her front feet braced on the back of the driver's seat. Pooch rode to all the fires, head high in the wind. She rode through sunshine, rain and snow.

Before Pooch came to live at the fire hall, she was a stray. Dirty and almost starved, she had come close to being killed several times as she tried to cross busy Charlotte Avenue.

After a bath and a good "fire hall meal" she adopted Engine 13 as her permanent home and the firemen as her best friends.

Her career as a lady firefighter began one day as the firemen ran for the engine to answer a dump fire call at Forty-second Avenue North and Delaware Avenue. One of them picked Pooch up and put her on the engine. From then on she was first on the engine to answer every call the engine made. It was a fact that she answered more calls than the firemen did because she was on duty twenty-four hours every day. The firemen worked twelve-hour shifts.

Pooch did miss two or three calls when she became the mother of a litter of six puppies. The firemen received fifty or sixty requests for a puppy as West Nashvillians expressed their love for the little brown and white fire hall mascot.

ROY

"Have you heard about what happened to Roy?" This question echoed throughout the Nashville Fire Department that Tuesday in 1915.

There is no description of Nashville's most beloved firehorse in the Fire Department records. Only that he was assigned to Wagon Team 1 and that he answered more fire calls than any other horse in the department.

After fifteen years of service, Roy fell and broke his leg while answering a 2:30 a.m. alarm at the Y.M.C.A.

It was with deep regret that Chief Rozetta ordered the big horse shot. Nashvillians also mourned the passing of their favorite firehorse, Roy.

4 THE LAST ALARM

The last alarm has sounded for these brave men of the Nashville Fire Department. They lost their lives in the faithful and conscientious discharge of duty and in heroic efforts to save the property of citizens of this City and County.

Such heroism should not go unnoticed or unrewarded. They forgot themselves as they battled for the interests and safety of others and gave up their lives in the heroic discharge of their duty as only brave men can do.

Irreparable loss has been brought upon their families and great loss to all citizens of this area.

We regret that an equal amount of information concerning each man was not available. We realize that each man gave his full measure. We shall ever hold them in honored remembrance.

C.C. GOWDY • J.H. EWING • S. ALLEN

Their Names Seared into City's History

Billy H. Easley • Focus Columnist • *The Tennessean*

Downtown-It was 96 years ago January when three black Nashville firefighters were immortalized by a tragic accident while fighting a fire on Third Avenue North.

Capt. Charles C. Gowdy, hoseman J. Harvey Ewing and reel driver Stokely Allen of Stockell Fire Co. 4 were killed when a brick wall collapsed on them as they directed a stream of water on a blazing inferno.

The unfortunate firemen were standing on the roof of the Phillips and Buttroff Co. building directing a stream of water onto the burning Weakley Warren Furniture Co. at 215 College St. (Third Avenue) when the store's north wall collapsed. It fell on the firemen who were on the roof of the building across the alley at 217 N. College.

The fire began on the seventh floor of the building occupied by the Webb Stevenson Co. 213 North College St. Engine Company 4 was among the units responding to the alarm sounded at 5:48 p.m. Jan. 2, 1892.

Reporters at the scene said, "Horse drawn firewagons from every section of the city rolled into Printers Alley. The smell of smoke was heavy in the air, but there was no sign of fire. The firemen, answering the alarm box, looked up at the row of buildings fronting on North College Street. Suddenly a pillar of fire shot from the roof to the Webb Stevenson Co. building."

News accounts quoted witnesses as saying Gowdy was holding the nozzle directing a stream of water on the wall and roof, and Ewing and Allen behind him, when the wall wavered and collapsed on them.

The accounts said high winds caused the fire to spread quickly and the bitter cold hindered firefighting efforts.

The five-hour Saturday night blaze wiped out an entire city block—eleven buildings valued at half a million dollars. It was the worst fire ever in Nashville's downtown business district.

The charred bodies of Allen and Ewing were not recovered until the next morning; Gowdy's body was found in the debris and ashes Monday with "his charred hand still clutching the nozzle of the hose."

A joint service was held for the three in the state Capitol rotunda attended by thousands. Schools were dismissed, courts adjourned and business houses closed during the Wednesday funeral.

Funeral Director Preston Taylor devised a special carriage for all three bodies to rest side by side for the processional from the Customs House to the Capitol.

Taylor, who was also a minister, said of the three, "They were brave, noble fellows and their untimely end and devotion shall not be forgotten."

Stockell Fire Co. No.4 was the first all-black firefighting unit in Nashville. The company was organized Jan. 15, 1885 and was stationed at 424 Woodland St.

Original members of the company included Gowdy, Lt. Aaron Cockrill, who became captain when Gowdy was killed, James Trimble, Ross Moore, John Harris, James Rogan, Charles Vaughn, Ewing and Allen.

Gowdy, 38, who was born in slavery, resigned a position on the City Council to take the fire department position. He had been a constable from 1878-84 and was a community leader. His wife, Mary, was a teacher at Knowles School. His son was 15 at the time of Gowdy's untimely death. The family lived at 423 High St. (Sixth Avenue).

Allen, Capt. Gowdy's brother-in-law, was 29. He left a young wife and small child.

J. Harvey Ewing's grandson, Capt. Fred Ewing, recently retired from the Metro Fire Department. He still has the nozzle which was found in Capt. Gowdy's hand nearly a century ago.

Gowdy was the highest ranking black fireman in the nation at the time of his death.

Newspaper accounts reported community resistance to the black company but later acceptance and admiration when the company proved themselves exceptionally capable.

The outpouring of sympathy for the dead firefighters apparently broke down all bias and all barriers, allowing the black unit—later Company No. 11—unquestioned acceptance until the day the department was fully integrated.

(Feb. 3, 1988 — Reprinted by permission of Tressa Wasson)

ROBERT E. PAGE

Robert E. Page, reel driver of the McLemore Street Engine Company, was killed by falling roof on November 11, 1898, at the University of Tennessee Medical Department fire on Broad Street.

At 6:42 o'clock an alarm from box No. 7 was turned in, and on responding to the call the department found the three-story building of the medical department of the University of Tennessee, located at 614 Broad Street, between High and Vine Streets, wrapped in flames. While fighting the fire Robert Page, reel driver, met his death; William Silvers of the North College Street Chemical Company was seriously injured; and John McCormick of Engine Company 3 sustained injuries which [would] incapacitate him from duty for several days. At the first alarm Company 3 of West Broad Street, #5 of McLemore Street, #6 of South Market Street, #3 Chemical Company of North College Street, and Hook and Ladder Truck of Broad Street were called out. As the fire promised to be of stubborn resistance, a telephone message to #2 summoned that company also to the scene of the conflagration. Captain Carell saw that the fire was more advanced in the upper story and ordered Company 5 to attach an extra hose to its engine, to be used on the third floor of the building. Driver Page of the McLemore reel, together with firemen Cohn and Connette of the same company and William Silvers were ordered to the top story of the building. They had been fighting from that vantage point for some little time, and the fire was thought to have been under control when an explosion of a barrel of alcohol caused the west wall of the building to fall. The falling wall served as a signal of danger to the men on third floor, and all of them made a rush for the ladder. All except Page and Silvers had just reached the place of safety when the heavy mansard roof covering the building caved in, catching Page and crushing him to death, and also wounding Silvers. The presence of mind of the wounded man was remarkable under the circumstances. In a half-dazed condition he banished all thoughts of personal safety, for after extracting himself from a perilous position, he rushed to the point where Page fell and attempted to drag him out of the debris. His efforts were fruitless, as heavy timbers held the body of the courageous fireman in their grasp. Cohn and Connette who were close at hand, also made a super-human effort to extricate their fellow fireman, but failed. The men of Company 2 of North Market Street took in the situation, and one after another began to hurriedly mount the ladder. When they reached the top floor they were blinded by smoke and dust from the falling wall. The voice of Silvers directed the rescuing corps to the spot where Page was pinned down. The fireman held on to Page's arm with one hand and with the other was trying to aid McCormick and Cohn in releasing the mangled and lifeless body of Page. The only part visible of the dead fireman were his head and hands. The members of Company 3 brought their axes into play and in short time released the body of their dead comrade.

WOODRING REPORT

CAPTAIN E. M. CARELL

(This appeared in the *Nashville Banner* March 6, 1900.)

Captain E. M. Carell, after days and nights of patient suffering, died this morning at 12:14 o'clock. At that hour death entered the chamber at St. Thomas Hospital and claimed the brave fireman. He had been hovering between life and death since last Wednesday morning and though the end was not unexpected, it is attended with sorrow and gloom such as it is not often manifested by the public at large. Years of faithful and efficient service had endeared him to the people at Nashville as his many acts of heroism and courage, displayed in hundreds of instances, had won their admiration. He was essentially a man of the people and the latter realizing this entertained for him the kindliest and most loving sentiments.

In the passing away of Captain Carell, Nashville loses one of its most efficient public servants. For thirty years, except for a brief period, he had been connected with the fire department. It may be said that he literally grew up with the department as he filled every position in the service except that of engineer. While working in the mechanical department of the Nashville, Chattanooga & St. Louis Railway, Captain Carell became a lantern boy in the old volunteer department. When the city established a paid department he was employed as a runner. This was in October 1870. He was ambitious and he rose gradually until in 1884 he was chosen to command the department.

Under his careful guidance the service was greatly improved; additional apparatus was purchased at his suggestion; discipline was systematized among the men, and at the time of his death Captain Carell had the satisfaction of knowing that Nashville has one of the best equipped departments in the country.

Although a man who took a deep interest in everything pertaining to the welfare of Nashville, Captain Carell kept himself aloof from politics and each succeeding administration kept him in the position. He was a strict disciplinarian and though he required his men to do their full duty, they respected and loved him. He would never ask them to go anywhere he did not go himself, and his commanding presence often led them into dangers at fires which thrilled spectators and caused them to fear for their safety. Captain Carell was, in fact, a stranger to fear and danger. It was this spirit which on more than one occasion caused him to be injured in the discharge of duty. Frequent warnings by friends went unheeded, as he seemed to be perfectly oblivious to self when duty beckoned.

Captain Carell received the injury which resulted in his death while responding to an alarm in East Nashville last Wednesday morning. While passing along the Public Square, his buggy collided with an electric car post which stands almost opposite the mouth of North Market Street. He and his driver, Charles Benson, were thrown out and Captain Carell received a fracture of the skull. Benson was only slightly injured and has about recovered. Captain Carell was conveyed in an unconscious condition to his home on North Summer street and was subsequently removed to St. Thomas Hospital, where his skull was trepanned. Although everything that service could suggest was brought into use, the injury proved fatal.

Fire Chief E.M. Carell's buggy #2, 1900. Driver unidentified.

ROBERT C. BURK

Tragedy struck about 10:00 a.m. on Thursday, September 10, 1903. The actual damage done by the fire which was the resultant cause of this horrible accident was insignificant as compared with the loss caused by the collapse of the buildings.

The fire broke out at 4:15 a.m. on the third floor of the Coleman, Tompkins and Company's wholesale grocery firm on Broad Street. The blaze was discovered by a night watchman as he made his rounds. He quickly turned in the alarm at Box 12. Every fire company in the Nashville Fire Department answered the call, which was customary when an alarm sounded from that area of town. The buildings located there were "some of the oldest in Nashville."

Firemen found heavy boxes were stacked up against and to the top of the windows.

"The probable cause of the blaze was rats playing with sulphur matches," Asst. Chief. J. W. Weaver said later. "They gave off an almost suffocating gas. I think we must have shoved out over a carload from the window in getting at the fire."

The flames quickly spread to the Broad Street Stove and Tinware Company next door to the west.

The blaze was stubborn. The roofs of both buildings burned. With heroic efforts of the firemen and thousands of gallons of water, the fire was contained to those two stores. The water flooded other stores to the east and west.

At 6:00 a.m. the fire was believed to be under control. Chief Weaver ordered all but three companies to return to quarters. Some of the firemen left on the scene were in the street taking up the hose, preparing to return to quarters.

Inside the building on the third floor, Chief Weaver was supervising the salvage and overhaul. A 2½" line had been advanced through the window with Hose Company 1. Captain Robert C. Burk, 49, was in immediate command. He had been an efficient fireman for fifteen years and a captain for eight years. He knew his job well and was proceeding with his men in the drowning out of hot spots.

Chief Weaver took precautions against collapse of the front wall. He did not forsee that both buildings, one on the east and one on the west, would collapse into the Coleman, Tompkins and Company. Apparently no one did.

Loud cries came from the street, "The walls are bulging!"

Suddenly the front wall of the Broad Street Tinware Company began to give way. A great rumbling was heard and, as if by one mighty impulse, the walls of the three stores came down.

John Yeaman, an ex-fireman, was the first man to reach the pile of debris. Others quickly joined him, digging frantically, guided by the groans of the trapped firemen.

The first to be reached was Fire Chief Joe Weaver. He was completely covered by bricks and heavy timbers. He could be heard calling, "For God's sake save us!" Although his head, legs and lower body were pinned in, the heavy timbers had fallen in such a way as to make a protecting cover which saved his life.

In spite of the heat and dust, rescuers pushed their way fifteen feet farther into the building to find Capt. Robert Burk. He, too, was completely covered with debris. Nearby Sam McHenry and George Farries also had to be dug from the fallen rubble. They had been working on the third floor, but had fallen down to the first floor when the buildings collapsed.

Captain Robert Burk was the most seriously injured, suffering a fractured spine and other internal injuries. Death came for the brave and efficient Captain of Engine Company 1 later that day. He was survived by a wife and eight children.

Funeral of Captain R.C. Burk, September 12, 1903.

GEORGE ROGERS

George Rogers was a six-foot blond young man of 22 in 1905. He possessed a sunny disposition that made him the life of every circle in which he moved.

George's father, Henry W. Rogers, was a former firefighter. Both his sons, W. D. (Dock) Rogers, 25, and George S. Rogers, 22, were substitute firemen.

On the fateful night of July 3, 1905, twelve men worked overtime at the Philips-Webb Company, a wholesale grocery firm located at 144-46 Second Avenue North. Behind on their orders because of the coming holiday, they left work around midnight.

At 2:00 a.m. two patrolmen, walking their beat, smelled smoke. They discoverd black smoke seeping under the door of the firm and immediately turned in a fire alarm. Engine Company 2 arrived first. When Chief Rozetta arrived, he quickly sounded a general alarm. The whole interior of the building was enveloped in thick black smoke, and it was evident that the entire structure would be a total loss.

Soon Engines Nos. 1, 2, 5 and 6 were fighting the fire from the front of the building. Hose Company 2 and Hose Company 1 were located on the roof of the H.G. Lipscomb & Company building. Chemical Engine Company 1 had a hose in the plant of the Cheek-Neal Coffee Company next door.

George Rogers was off duty. He did not have to go, but he knew there was a fire and that his brother was fighting it. He walked from his home at 209 Trutland Street across the river to the square. Pushing his way through the stifling black smoke, the crowd, and the equipment, he kept asking where Engine 2's men were located. When he finally found his brother, George held the lantern for Dock. They were moving very slowly. All of a sudden George yelled "Look out!" then dropped out of sight in his tracks. George had fallen down an elevator shaft. He landed astride a metal grid erected around the elevator car to keep freight from falling off. When fellow firefighters got to him, he was still conscious but mortally injured. His flesh was badly torn, he had suffered a broken leg and an artery had been severed. He was losing blood fast.

Death came July 6, at 11:35 a.m. Since he was a substitute fireman there was no black-draped engine for George as there would have been for a full-fledged company member, but firemen gathered at Mt. Olivet Cemetery to honor one of their own.

Dock Rogers remained with the fire department for some time, but later joined the Tennessee Inspection Bureau as an instructor in better and safer methods of fire fighting. He remained in this capacity for 35 years.

He always said, with tears in his eyes, "Yes, George saved me!"

WILLIAM R. GREEN

It's hard to believe today that a kite string, wires and a rotten telephone pole would be instrumental in a fireman's death. But in 1909 the city's fire alarm system had 500 miles of lines within the city. Eighty per cent of these were strung on telephone poles. For 35 years the fire department had maintained its own lines.

On November 3, 1909, William R. Green and two workers who were brothers, R. M. Cassetty and Frank D. Cassetty, left old Headquarters on Fifth Avenue in a horse-drawn wagon loaded with new wire and tools. The three workmen clattered along to their destination at Edgehill and Eleventh Avenues. At that corner an old pole was being replaced by a new one. The old pole was still standing. Part of the wires were down, entangled in kite string. Other wires were intact on the cross bars. All had to be removed and replaced onto the new pole.

The Cassetty brothers aided from the ground while Green, wearing spikes, climbed the pole. When he had gotten 15 or 20 feet up the pole, they all heard a loud snap. The pole, being rotten at the ground, had broken off. Green realized his danger and cried out. He could not disengage his spikes in time to jump. He had to ride the pole down. R. M. Cassetty attempted to guide the pole, even though the wires cut deep gashes in his hands. It seemed that he was succeeding, but when the pole crashed on the sidewalk it bounced and dropped back onto Green.

Death came instantly. His skull was crushed and his right hip and knee were crushed. The Cassettys called for help but it was in vain.

The Cassettys worked on at the alarm station. The family became a tradition where several members served a total of more than 77 years.

FALLING WALL BURIES FIREMEN
LIEUTENANT BUD AMBROSE KILLED

(Woodring Report)

April 17, 1910—Lieutenant John C. (Bud) Ambrose, for many years a member of the Nashville Fire Department, was instantly killed yesterday morning about 8 o'clock by a falling brick wall in a fire at the warehouse of the Bradford Wholesale Furniture Co., near the junction of Jo Johnson Avenue and the N. C. & St. L. Ry. Four of his fellow firemen were injured.

Arch B. McKinley, of Hook and Ladder Company 2, suffered a broken leg.

Ed R. Norris, of Hose Company 2, received a cut in the forehead.

Capt. Frank Hite, of Hook and Ladder Company 2, is suffering with a sprained back.

W. A. Coggins, pipeman of Hose Company 2, received scalp wounds but was able to continue on duty.

The fire was said to have been caused by a stroke of lightning which, it is thought, entered the building over an electric wire.

About 2 o'clock a terrific thunder and lightning storm set in. Rain fell in torrents. The wind was high. John Spieler, night watchman at the plant, was lying on his cot. He was awakened by a flash of light through the building. It was a bolt of lightning, transmitted through the meter box, into the building. Immediately, the watchman said, the north side of the second floor was ablaze. Running from the building he yelled "Fire!" to John Bell, a driver for the company, who lived nearby. A district alarm was turned in from the plant of the American Snuff Company. Engine Companies 5 and 7, Hose Company 2 and Truck 2 responded.

Captain W.T. Climer of Engine Company 7, the first to arrive on the scene, said that the second floor of the building was then a mass of flames. Despite the heavy downpour of rain, long tongues of fire were shooting upward. An alarm was turned in. When Chief Rozetta arrived he had headquarters and Engine Company 8 to report.

The men had engaged themselves in cooling the walls and quenching the live coals among the ruins, when the accident happened in which one of them lost his life.

The north wall had already fallen in. It crumbled down about 7:30 o'clock. The men were inside the enclosure directing their attention to the south side. Besides Lieutenant Ambrose and the four injured men, Captain Climer of Truck 2 and James Andrews of Hose Company 2 were standing in the shadow of the wall. They were directing two streams on the flames that occasionally burst forth from the ruins.

Lieutenant Ambrose was holding the nozzle. He was shooting a stream into the debris when the wall began to crumble. Captain Climer, standing a little to one side, saw the wall sway toward them and gave the alarm. But it was too late. Tons of the thirteen-inch brick wall came down in a mass. Six of the men were buried under the debris. Captain Climer escaped by rare good fortune. Pipeman James Andrews was knocked to his knees.

Just as the wall caved in Captain Weaver, who was relieving Chief Rozetta, drove up. He had the remaining firemen begin the rescue. Captain Frank Hite was caught from his waist down. He was the first rescued. Ladderman Arch McKinley was lying on his face in the thick of the debris. His right leg had been caught up and doubled back. Both bones were broken between the knee and ankle. Only three or four feet away they found Lieutenant Ambrose. His body was entirely hidden under the ruins. When he was first uncovered no injuries could be seen except a cut under the chin and a bruise about the right eye. But he was unconscious. His neck was broken. He breathed only twice after being rescued. Ed Norris and Pipeman Coggins were only partially buried. The former received a cut in the forehead and the latter a scalp wound.

"We scarcely had a warning of the coming calamity," said Pipeman Coggins, one of the men who carried down the injured. "We were all standing there fighting the fire, as we had been for six hours, when one of the fellows shouted that the wall was falling. I looked up and saw it. A great sheet of brick was coming down in a mass. I jumped. Before I could clear myself I was caught and knocked to my knees. The bricks fell all around me like so many raindrops."

Lieutenant Ambrose was a member of the Fire Department for eighteen years. He was promoted to Lieutenant fifteen years ago. He had served in West Nashville and in North Nashville when the present Chief was Captain of the first company. At the time of his death he was under Captain Hyde, at the Jefferson Company. He was 52 years old. He was known not only as a good fireman, but as a good fellow among the firemen. He was also known to all of them as the "fish fryer" of the Fire Department. He was always invited to every fish fry and prepared the fish in a delicious manner. For many years he was a member of the Fireman's Benefit Association and served as director of the association.

Chief Rozetta recognized Lieutenant Ambrose as one of the very best members of the department. He was especially liked among the members of his own company, Hose 2. Captain Hyde, his senior officer, said of him yesterday: "In the death of Lieutenant Ambrose we have lost not only a fireman of the first rate, but a good man."

WALLACE PINKERTON

At 7:00 a.m. on June 1, 1918, Engine Company 7 was speeding north on 18th Avenue North to answer a call at Box 142 at 18th Avenue North and Cedar Street.

Sub-driver Alex McWright was driving. Lieutenant Wallace Pinkerton sat on the seat beside him, cranking a hand siren.

Standing on the rear platform of the Engine was Pipeman John Barksdale and Pipeman Earl P. Brown. Brown was ringing the bell.

They all saw a street car outbound on Church Street speeding toward them. There was a stop-ship system in effect, but they all expected the motorman to hear the siren and bell and stop at the corner. It was later learned that the motorman was talking to a mail carrier and did not see the approaching fire engine until it was too late. On seeing the impending accident, the motorman applied his air brakes but the street car slid into the rear of the engine.

McWright, driver of the engine, had also seen that an accident was unavoidable and at first slowed the engine, but then speeded up in an effort to get across the street car tracks. The impact was so terrific that the engine traveled 30 feet before the rear of the engine crashed into a tree. The engine was turned completely around. The front of the engine was intact but the rear was completely demolished.

The firemen were hurled into the air. Alex McWright suffered internal injuries. John Barksdale's head was gashed and he was bruised. Earl Brown received a broken ankle and facial cuts and was badly bruised. Lieutenant Wallace Pinkerton suffered a fractured skull when his head hit the curbing. His injuries were so severe that death came at 9:00 a.m. at St. Thomas Hospital.

CHIEF ROZETTA'S LAST RUN

"He came to America as a wee boy with a bit of the azure of his native Italian sky in his eyes, and something of Italy's sunshine in his face. The music of Italia was in his soul, the strength and courage of old Rome was in his heart!" eulogized Dr. Carey E. Morgan, Minister of Vine Street Christian Church, as he stood beside the gray, flower-banked casket of Nashville's Fire Chief A. A. Rozetta.

At 2:40 p.m. that Saturday saddened firemen placed his casket aboard the gleaming red fire-fighting machine, Headquarters Engine 2, that had only a few weeks before been christened "Chief A. A. Rozetta."

It was fitting that the great new engine should serve as a warrior's caisson, because the undertaker had reported that the body of the grand old soldier was covered with scars received in his many battles with his greatest enemy — fire.

The procession the Chief led on that day was not the thundering feet of the gallant firehorses with bells clanging, or the roar of the red engines with lights flashing and siren blowing, as he had led them countless times during his twenty years as Chief of the Nashville Fire Department.

No, this was a slow funeral march, as firemen with tear-stained faces marched or rode behind their beloved Chief. All engine companies that could be spared followed: Numbers 1, 2, 4, 5, 6, 7, 8, 9 and 10.

During the twenty years he served as Chief, many innovations and improvements were made. The most sweeping change began in 1900 and continued for eight years—the change from horse-drawn engines to motor vehicles. This was not an easy task for Chief Rozetta. His love for the gallant steeds had long been a tradition for firemen and the general public of Nashville. A strange party was given April 27, 1920, at Headquarters on Fifth Avenue. It was given to christen Nashville's newest and finest fire engine, "Chief A. A. Rozetta." Although the firehall was decorated with American flags and bright Chinese lanterns, the crowd was not happy—something was missing. No friendly whines and no clump of horses' feet could be heard—the horse stalls out back of the firehall were empty.

Many dignitaries spoke in tender tribute of the Chief's love for his work, his thirty-six years of faithful service to the city, his love for his men and how he appreciated "everything helpful and hopeful."

Now as the dear old Chief rose, everyone expected him to speak of progress. He did not. His speech, entitled "The Passing of the Horses", brought tears to the eyes of those present. He stood erect and as he called the names of the faithful old horses he paused, as if to recall a personal friendship and pay a final tribute: Fite— Pink— Dude— Tom— Jere— Mun C.— Morris W.— Warren— Chris— Billie P.— Judge— Benson— Leuis— Buck— Push— Wilderson— Morris C.— Niger— Ned— Lee— Charlie— Henry— Dash— Dan— Hilary— Skeet— Gyp— Henry— Bulger— Mack— Bob Taylor— Bob— Frank— China— Cherry— Pete— Ben— Max— Prince— Bart— Matt— Pat.

Then, in closing, he read the following poem:

Then he rubbed down its nickeled and varnished coat,
And he shined up its great glass eye,
He polished the brass with a lump in his throat,
And a sorrowful, long-drawn sigh,
He lifted the hood where its metal soul
Lay hidden and all unseen,
Then unscrewed a cap from a yawning hole,
And fed it some gasoline.

Everyone recognized that the motor vehicles were faster and more economical, but their cold impersonal efficiency could never replace the love and bravery of the firehorse. Thus, the passing of an era in the Fire Department.

A few weeks after this party, death came at 9:25, July 21, 1920. In the funeral service, Dr. Carey E. Morgan made reference to the East Nashville Fire—"I shall never forget how I found Chief Rozetta on the late afternoon of that fatal day or how he looked. No commander on a field of battle ever strove harder to win a victory, no soldier ever fought more valiantly than he. He was stained over from head to foot as if with battle scars—was spent and worn and weary but still courageous. Perhaps in that terrible struggle, the seeds of death were planted in his system."

THURMOND OSCAR DILLARD

Life slipped away from Thurmond Oscar Dillard at 5:45, September 17, 1921, at the Dozier Clinic at 8th Avenue and Monroe Street in Nashville, Tennessee. It all happened because, as Chief J. W. Weaver said, "His over anxiety to do his duty cost Dilliard his life — he was one of the most daring and brave men on the fire force."

T. O. (his nickname) and "Little Kit" (T. O.'s nickname for his wife) and their small son lived at 523 North First Street. For the three years he had been on the fire department he had walked the four blocks to the firehall on Foster Street where he worked on an old engine named "Stainback."

On that Friday, T. O., "Little Kit", and their little son had enjoyed a picnic with relatives and friends. At 5:30, two hours before he normally went to work, he asked his wife and son to walk to the gate to kiss him bye. He wanted to relieve Engineer J. I. Ferrell early that day. Little Kit and the boy watched as T. O. walked down North First Street and turned on Foster Street.

One of the neighborhood boys, Russell Campbell, who idolized firemen, rode his bike beside T. O. and said, "Let's race!"

T. O. said, "It's too wet. We'll race another time."

Suddenly they heard the siren, the bells and the roar (fire engines at that time did not have a muffler) of the old "Stainback" coming west on Foster Street. T. O. decided to run back to the corner of North First Street, so that when the engine turned the corner he could hop aboard. At the very moment he jumped, the engine skidded on the wet pavement. The ladder hit him on the chin and knocked him down. The rear wheel passed diagonally over his body.

The two firemen on the platform saw what happened; but because of the noise of the engine, they could not make the driver hear them.

The boy on the bike rushed to tell "Little Kit." When he got to her she had already guessed what had happened. "Was that T. O.?" she asked.

Russell said, "Yes ma'm."

All of this caused by a $10.00 fire in a closet—a man lost his life, Little Kit became a widow, and a little boy became fatherless.

Evidently this incident caused Russell Campbell to pursue his ambition to become a fireman. He joined the department in 1930. He advanced through the ranks to become Chief in 1963.

JOHN S. (PENNY) PENNINGTON

John worked four years at Engine Company 7 at 1712 Broad Street. On his service record was written: *Valiant, efficient and fearless.* His wife, Mary, tells that he was merry hearted and loved children, animals and everything else that God made.

On March 27, 1922, John put on his new uniform that Mary had given him just three days earlier. It had been a surprise. John had told Mary several weeks before that he desperately needed a new uniform for an upcoming event of the fire department, but he knew they could not spare money for a uniform. They needed food and clothes for Henry who was in school and little John Jr. who was just a baby. Mary had been sick. There were high doctor bills that needed to be paid. His salary of $100.00 a month just couldn't be stretched to include a new uniform.

Mary smiled to herself. She had anticipated this. She had saved some money in an old black sock. She knew John's measurements. She ordered the uniform.

On that fateful day, at 11 a.m., the alarm sounded from Flynn's Garage at 15th and old Cedar Street (now Charlotte Avenue). Engine 7 and Engine 11 from 1004 Jefferson Street responded. Engine 7 arrived first on the scene. The building was blazing. John grabbed the hose and ran to the fire hydrant. Engine 11 pulled into 15th right behind Engine 7. The driver of Engine 11's view was obstructed by a huge sign. He did not see John. The engine struck John, breaking one leg and several ribs. John was taken to old Protestant Hospital. In delirium John pleaded, "Don't tell my wife. She's been sick." After further examination it was found that John had a severely fractured skull.

Death came that afternoon. Mary buried John in his new uniform.

Life was not easy for Mary after she buried her husband. She earned money doing jobs like keeping children and stretching curtains. Circumstances got so bad she was afraid to appeal to organized charity, fearing the authorities would take her son away from her.

JOHNSON VAN TIEN

Johnson Van Tien had one ambition all of his life. "I'm going to be a fireman before I die," he said as a schoolboy. He did achieve his ambition when he was twenty-five years old. Short, stocky, dark haired, light hearted, popular, and generous to a fault, he often gave his salary away. In fact, he had the reputation of always being good for a "touch."

By noon on Wednesday, June 28, 1933, the mercury had climbed to 92 degrees. The people of Middle Tennessee prayed for an end to the drought that had gripped the area for some time. That afternoon there were shouts of joy as the rain came pouring down, the lightning flashed and the thunder rolled.

Johnson Van Tien rushed around the fire hall to get his duties done as the "ice man" for that day. He knew that lightning could very well mean a fire call. He was right. The alarm sounded and his engine,

"Major E. B. Stahlman," roared out of Engine Company 17 on Fairfax Avenue. Driving was Engineer Johnny Miller. Captain H. D. Demonbreun sat beside the driver. On the back platform rode Jesse K. Vaughn and Van Tien.

That run did not take long, and they returned to the firehall with the assistant engineer, Jesse Vaughn, driving. While he was backing the engine into the hall, they received another call. This time lightning had struck the vent on top of a storage tank at an air reduction company, where tanks were filled for acetylene welding.

The "Stahlman" sped past the fire hall at about 35 miles per hour. A lady motorist pulled into the street and drove just ahead of the engine. Swerving sharply to avoid hitting her car, Vaughn was now trying to cope with rain-slick streets and wet slippery car tracks. The engine went into a spin that carried it sliding 100 yards and up on the sidewalk into a utility pole. When the engine stopped spinning it was headed in the opposite direction.

Van Tien was thrown from the platform. Miller was thrown over the engine and landed 65 feet away. He was to lie in a coma for seven days, suffering from an injured spine and a shattered collarbone. Demonbreun was thrown from the engine but not hurt. Vaughn held to the wheel and was not hurt.

Van Tien lay near the left wheel of the engine with cuts on his face and legs. His skull was fractured and his legs were crushed. Rushed to St. Thomas Hospital, he died while being wheeled to the operating room. Van Tien gave his life while fulfilling his ambition—"I'm going to be a fireman before I die."

W. H. NORRIS
F. R. "BUDDY" REECE
TRUMAN M. HILL

A freak accident claimed the lives of three of Nashville's finest firemen on February 6, 1934. Fire Chief Frank B. Moore termed the wreck the "worst accident since I joined the force in 1892."

At 10:35 Engine 7 left her fire hall at 1712 Broad Street. She was thundering down Broad headed to 18th and Jo Johnson. There were five firemen and a "stowaway" making that fateful run. Captain William "Will" H. Norris, 29, who had the reputation of "having what it takes to one day be chief," was in charge. A.A. Isaacs, 29, was riding in the metal "basket." Buddy Reece, the youngest of the firemen, had been a fireman for only 3½ years. The father of three, he and his wife, Lena, were expecting another child soon. Engineer Truman M. Hill, a fifteen-year-veteran firefighter, was known as the driver who "had never knocked a piece of paint off his engine."

L. O. Amos, 34, and Reece and the "stowaway" were riding on the platform.

A milk truck from the West Lawn Dairy, driving west on Patterson Street, pulled into the intersection.

With siren blaring, the engine was thundering toward its destination. Hill swerved and the front of the engine missed the front of the milk truck, but the rear of the engine struck the bumper and fender. As Hill

Courtesy of Metro Nashville Archives

Engine Company #7 struck by milk truck at 18th and Patterson, killing three firemen, February 1934.

tried to pull back across the center line, the engine was thrown off balance. It careened across the intersection, momentarily "straightening out," but the 8,000-pound engine was off balance. The noise was awful as the huge machine turned completely upside down from the force of the swerve and slid sideways several feet. The engine and all equipment were completely demolished.

The "stowaway" was unhurt. He ran to call for help. Hill and Reece died instantly. Norris lived until 4:40 that day. He had received a fractured skull, a broken arm and leg. He died at St. Thomas Hospital.

Amos and Isaacs later drew pensions as a result of their injuries.

An editorial dedicated to these men read: "There are enough hazards in the life of a fireman without the inexcusable hazard of having to sacrifice their lives in preference to killing a trespasser who is blocking their way to the battlefield."

JOE KEEF • T. MADDEN • C. OBENCHAIN
W.E. Stephens Manufacturing Company Fire

Heroism and tragedy marked the work of the firefighters who fought the fire at W. E. Stephens Manufacturing Company on the Public Square, February 20, 1937.

It was dark and cold as the first inbound Inglewood bus rumbled across the Woodland Street bridge. On the bus was T. R. Rogers headed for his work as a postman. As the bus crossed the bridge he saw the red glow of fire in the basement section of the Stephens Manufacturing Company. Rogers jumped off the bus and raced for the nearest fire alarm box to report the fire. This fire alarm was turned in at 5:34 a.m, a second at 5:42, a third a 5:47 and the fourth at 6:23. Later Chief Frank B. Moore called it the "biggest fire since the East Nashville fire in 1916."

Eleven engines and four truck companies were engaged in fighting the fire. The engine companies were Nos. 1, 2, 3, 4, 5, 6, 7, 8, 10, 12 and 14. The truck companies were Nos. 1, 2 3, and 4. A total of 60 firemen fought the fire.

The fire presumably started from a small coal stove that had been left burning in the basement during the night.

The 100-year-old timbers and cotton fabric for the clothing manufactured inside the three-story brick building provided a ready fuel supply for the flames. Evidently the fire had been burning for some time because when firemen arrived on the scene, flames were leaping wildly inside the building.

The building backed up to the river bluff. On one side was the Neeley Harwell Company building, on the other side across the alley was the A. L. Kornman Company building. All three buildings were similar in structure and were about 100 years old.

Because of this location, firemen were forced to fight the fire only from the front of the building. The origin of the blaze was in the rear of the building in the basement near the river bluff. Firemen had great difficulty stretching hoses to that area of the building because of intense heat and heavy smoke.

For four and a half hours the sixty firemen battled the searing heat and black dense smoke. There were times when it appeared that the whole block would burn. Three times Lieutenant Joe Keef on Engine Company 1 appealed to Chief Moore to let him take his company to fight the fire "from underneath." Twice Chief Moore refused to let them go into the building. On the third request Joe Keef said "Chief, I believe I can go in there and stop it." Chief Moore relented. Eight men, led by Keef, lugged a huge hose into the building. An old newspaper article gives this account of the impending tragedy: "Clouds of steam boiled up vanishing in the heat, as water began cascading toward the basement. The huge timbers of the ancient building began to burn a little less fiercely. The fuel was being choked off a little at the base. From windows and roof, great flags of fire still waved, fed the burning mass on the lower floor."

Firemen were gaining a small edge. Once the fire below was out, they could work more swiftly upward. Then came a thundering noise like the report of heavy gunfire. A wall of fire swept down.

The third floor already bore the wreckage of the roof, and the water soaked mass had collapsed upon the second floor. One by one in a boiling cascade of flame and bricks and charred timbers, the floors were falling — third, second, and first. Tons of debris were crashing on through into the basement, taking along drygood machinery and firemen. As it went two walls crashed forward. A great gush of sparks and flames shot toward the heavens and then sank back, smothering much of itself in its sudden new compactness.

After the roar died away all firemen were accounted for except Joe Keef, Thomas Madden and Charles Obenchain, all from Engine Company 1. Where were they? They did not answer. Were they buried beneath tons of debris? If so, could they survive? Comrades immediately started a rescue operation that was to continue for 16 grueling hours.

As a cold bitter night came on, lights were strung around the gruesome scene. Wind howled through the black shell. The side walls swayed and trembled, threatening fifty firemen frantically digging to find the bodies of their friends. Some used picks and shovels, but most knelt on the tons of blackened debris to dig with their hands. Wheelbarrows and tubs were used to carry debris to the alley. It was an ominous sight to see the six ambulances stand with stretchers in the street just waiting outside the roped-off area. Grief-stricken family and friends gathered with thousands of Nashvillians, dazed and shocked, hoping against hope that by some miracle the men would be found alive.

While the rescue operation was in progress, at General Hospital seven firemen were being treated for injuries sustained at the fire.

Dan Searcy, 1018 West Eastland Avenue, Truck Company 1, overcome by smoke.

Earl Turner, 301 Church Street, tiller on Truck Company 2, cut on left hand.

H. L. Burgess, 1608 Arthur Avenue, pipeman at Engine Company 1, overcome by smoke.

Shad Sullins, 1210 Fifth Avenue North, Engine Company 1, dislocated hip.

Curtis Buck, 1607 Fatherland Street, Engine Company 1, overcome by smoke.

N. H. Williams, 200 Seventh Avenue South, tiller on Truck Company 2, broken left arm.

E. L. Gullett, 4984 Nevada Avenue, Engine Company 12, twisted left knee.

As the hours bore on, hope drained from the grimy, sweat-streaked faces of the firemen. One said, "Oh, we'll find them all right but they will all be dead."

At 11:35 p.m. a cry rang out, "They have found one." The huge crowd surged forward as six firemen carried the limp body of Joe Keef from the pit of wrecked timbers and bricks. His body had been found under 8 feet of debris about 15 feet from the front door.

Just before Keef's body was brought to the surface, a cold rain began to fall. Because of the high wind, the swaying walls, and now the cold rain, the city building inspectors ordered search for the other bodies be halted until dawn.

At dawn firemen resumed their grim heart-breaking search. It was not until 12:07 noon Sunday that Tom Madden's body was found, near where Keef's body had been found. A foot away Obenchain lay face up, his left arm thrown across the body of his fellow fireman.

Joe Keef, 52, had been with the Fire Department about 25 years, a lieutenant for 5 years. He was survived by his widow, Hazel, one daughter and five sons.

Charles Obenchain was not married but was engaged and planned to marry soon. Survived by his elderly mother, he had been with the Fire Department for 5 years. He reported to work that fateful day even though he was scheduled to be off duty.

Thomas Madden, 40, was survived by his ailing wife, Ethel Madden, and a ten-year-old daughter he had nicknamed "Dolly."

"Bravery caused the death of the men," firemen said. "They didn't stop to think of the consequences."

"These men have fallen in line of duty," said an editorial, "as surely as soldiers on the battlefield. In the face of danger which caused superiors to order them back more than once, they firmly advanced into the very jaws of death, intent upon carrying out the traditions of the service. Their names will not lack for honor . . ."

BEN CHRISTIAN

Ben Christian made good. He was a conscientious, hard-working fireman, superior officers recalled. He worked on Engine Company 11 at 1204 Jefferson Street. Hc became a regular January 16, 1933.

On Friday, January 27, 1939, Ben worked on night shift. An alarm came in at 10 p.m. or later. Engine 11 sped to a house fire on Golf Street just off 12th Avenue.

The fire was soon out. Lieutenant Robert Campbell ordered the water shut off. Ben ran to the plug at the southwest corner of 12th and Jo Johnson to "shut it down." He began rolling up the hose. A truck came barreling out of 12th, hit Ben and sped away. Ben lay groaning in the street with a crushed thigh and chest, hip dislocated, his back severely injured and grave internal injuries.

Ben was taken to General Hospital where he groped in and out of consciousness. His injuries were too severe for recovery. His life came to an end at 4:30 a.m. January 31, 1939.

Since this accident, Nashville Police Department has provided protection for firemen at fire scenes.

ED DALE

Fireman Ed Dale suffered a heart attack in 1940. He worked at Engine Company 8.

LOUIS R. RYMAN
MAX SOLINSKI
S. B.(SHAD) SULLINS

Louis R. Ryman, 67, was a stout man with a big job which he did well. He was the engineer for the ''Fehr'' at Firehall Number 1 at 1312 Third Avenue North. He had a flawless record plus a family tradition to uphold, having married the daughter of the late Chief A. A. Rozetta. He was the father of six children and grandfather of eight. All the family were looking forward to the Christmas holidays. Ryman had come to work a few minutes early that Wednesday morning December 23, 1942.

The night men were preparing to go home and get ready for Christmas.

Another day shift man, Pipeman Max Solinski, arrived. He had been a regular member of the fire department for 12 years. An alarm came in from the King Clothing Store on Deadrick Street near Third Avenue.

''I'll take over,'' Ryman said to the man he was to relieve, and he mounted to the driver's seat for his last run.

Lieutenant Howard Hoffman climbed up beside Ryman and turned the siren as the big engine headed south on Third Avenue.

Shelah B. (Shad) Sullins, 56 and a veteran with 17 years service, stepped on the left side of the rear platform.

Pipeman George H. (Hub) Stephens, 32, was in the center position with Solinski standing on his right.

It was 6:30 a.m. but still dark. A swirling thick gray fog engulfed the mighty 12-cylinder engine as it thundered through the lowlands near the stockyards.

A troop train was approaching the grade crossing at Third Avenue between Harrison and Locust Streets. It was bound for Union Station.

Stephens remembered seeing a big cattle truck parked near the signal post, which could have obstructed Ryman's view.

The engine slowed a moment for the railroad crossing but suddenly Ryman saw the troop train coming at a high rate of speed. There was nothing to do but "shove her to the floor board." The front of the engine barely cleared the locomotive, but the locomotive struck the engine near the rear wheel. It sent the mighty engine "spinning like a top." The front of the engine snapped a utility pole. The bodies of the firemen were hurled through the air.

The engine halted upright 60 feet from impact. Lieutenant Hoffman lay on the floor board. His head had shattered the windshield, but his fire helmet had saved his life.

Solinski lay lifeless in the street, face up. The back of his head had been struck by the locomotive.

Nearby, Sullins lay very still. His face had borne the terrible impact.

Stephens had been hurled into the air. He suffered compound fractures in his right arm, spinal injuries, and an eardrum had burst.

Ryman was hurled to the street with his chest crushed, ribs broken, and grave internal injuries. He struggled for life in General Hospital but death came at 1:00 a.m. December 31.

Two men survived that terrible crash. Lieutenant Hoffman retired. Stephens was off duty five years but was able to return to work in 1948.

The "Fehr" would never make another run. It was so badly damaged that it was cut down and made into a tractor.

HORACE OVERHOLT

Overholt, 49, was stricken March 15, 1943, shortly after returning from a blaze. A second alarm came in. He asked for a substitute. Upon the company's return, Overholt, 12 years a fireman, was found dead in his firehall bed.

RAYMOND GOODRICH

Engineer Raymond Goodrich passed away January 1, 1944. He was an engineer at Engine #15.

JOE KNIGHT

Joe Knight, a man with a habit of hard work and efficiency, became a fireman in July, 1924, when he found a note on his door, "Report for work at the Fire Department." He became Lieutenant in 1933 after having worked on Engines 17, 8, and 7.

In 1945, after 21 years of service, his Engine 7 answered a call on 22nd Avenue South.

A huge transport truck loaded with 3,000 gallons of "airplane gas" had veered into a telephone pole, turned over, and hit an automobile before sliding to a halt upside down with gasoline spewing into the street.

As fire engines arrived on the scene, gas was gushing and fire was flashing as gas washed half a block down the street. Joe's Engine 7 arrived first on the scene. Three men grabbed the hose — Joe, C. L. Murphy and Jesse Scruggs. The situation was desperate, with thousands of gallons of gasoline spewing from the truck. Heat was so intense that houses along the street began to scorch and smoke, the brick sidewalk began to crumble and a huge tire on the transport exploded.

Firemen had to fill their boots with water as they blistered, stifling fumes swirled about them and black smoke engulfed them. Firemen could hear boiling noises inside the tanker.

"I'm sick!" gasped Joe Knight. He was helped to a porch where he became unconscious.

At General Hospital his condition, heat exhaustion and overcome by smoke, was not considered serious. He remained in the hospital overnight. As he was leaving the hospital his wife was assisting him. When he got to the car he collapsed onto the seat. He whispered, "I'm gone!" He had slipped away; it was 2:25, June 27, 1945. Poisonous fumes had attacked his lungs.

"He was one of the best men in the department," said Chief H. D. Demonbreun.

BERNARD MATTINGLY

Assistant Fire Chief Bernard S. "Bernie" Mattingly, 66, died shortly before 10 o'clock December 29, 1945. The veteran fireman was rushed to General Hospital after he did not respond to oxygen treatment at a fire.

In a desperate effort to save his assistant's life, Chief H. D. Demonbruen rushed to the fire department for an iron lung. Before the equipment arrived Bernie Mattingly had died.

Mattingly was a veteran of many of the worst fires in Nashville's history, including the East Nashville fire in 1916. He became a substitute fireman in 1914 and advanced in rank rapidly. A lieutenant captain before being made a battalion chief in 1939, he was named second assistant Chief in 1943.

Chief Mattingly was very popular throughout the Nashville Fire Department. Chief H. D. Demonbruen still dazed by his sudden death declared, "We have lost a very fine man. He was always a big help to the department."

WYLIE M. CARELL

Wylie M. Carell was 49 in 1945. He was the grandson of Chief E. M. Carell. He had worked at nearly every firehall in Nashville in the 30 years that he had been a fireman.

On this fateful day he was working at Engine Company 17 at Fairfax and Harding Road. The incident that occurred on December 29, 1945, would convince anyone that a false alarm or an erroneous call to the fire department is a very unwise thing.

Five firemen and three youthful occupants of a car were injured when a fire truck from Engine Company 17 and the car collided at Charlotte Road and 46th Avenue North. As a result of this accident Fireman Wylie Carell lost his life.

The fire truck from Engine Company 17 was answering a call to the Elite Theater on Charlotte, presumably in error. Clyde Spencer, manager of the theater, said that Fire Company 13 had been requested to send a ladder to replace a bulb on a pole in the parking lot next to the theater, but apparently a mistake was made.

Two other companies rushed trucks to the scene. As Engine Company 17 attempted to turn off of 46th onto Charlotte, parked cars made the turn difficult. The youth's car was traveling west on Charlotte but did not stop. The truck plowed into the side of the car. The impact was so violent that the engine did a complete about face and rolled, stopping on the sidewalk. All five firemen were thrown to the street.

The driver, Wylie Carell, and Captain W. B. Smith were on the front seat; both were thrown over the hood of the engine. Carell was unconscious at the scene of the wreck. He struggled for life for several days, but death came at 4:15 a.m. on January 3, 1946.

ORVILLE HOOBERRY

Fireman Orville Hooberry suffered a fatal heart attack at a fire on March 13, 1962.

J. H. COOPER

Engineer J.H. Cooper worked with Engine 2. He died April 19, 1954, after suffering a heart attack while driving home immediately after fighting a fire at Third and Franklin.

RAYMOND PARRISH

Engineer Raymond Parrish was a fireman for nearly twenty-four years. He became an Asst. Engineer in 1944 and Engineer in 1947.

He suffered a heart attack at Firehall 15 on April 25, 1956.

CHARLES COE

Charles Coe reported to work March 21, 1957, on Truck Company 2 on Fourth Avenue South. About 9:30 a.m. a call came from Kroger Grocery warehouse off Thompson Lane. Sparks from a welder's torch had fallen on a pile of plastic insulation board. The flames shot upward igniting a false ceiling. This was the beginning of a fire that would claim the life of Charles Coe and cause $100,000 damage to Kroger Company warehouse.

Within one hour flames broke through the roof. Near the rear of the building was a 1,000 gallon ammonia tank. It burst into flames sending a fog-like cloud of stifling fumes.

As the fumes engulfed him, Coe turned gray and became nauseated. He retreated to the fire's edge for fresh air but soon returned to fight the fire. After the fire was out, he and his company remained at the scene to drown out hot spots. He left at 5:30 a.m.

When he arrived home his wife knew something was wrong. He had never come home before in his firefighting coat with his face and hair coated in soot. He lay down on the floor, shivering from a chill and spitting up black mucus. "I can't get my breath very well," he said.

Coe lost consciousness then. The doctor pronounced the cause of death "breathing toxic fumes."

GEORGE H. "HUB" STEPHENS

George H. "Hub" Stephens loved his work with the Fire Department. It proved to be a very hard life for him.

On December 23, 1942, he was riding on the rear platform of Engine 1 when a troop train collided with the rear of the engine. This terrible accident claimed the lives of three of his fire fighting comrades. He sustained injuries to his arm and shoulder that kept him off duty for two years. He was overjoyed when he was able to return to duty.

On January 13, 1959, he took vacation to be at the hospital with his wife, Grace, who was to undergo major surgery. She was able for him to return to work after two days.

About 9:30 o'clock that very day, "Hub" suffered a fatal heart attack at a fire on Sylvan Street.

He was survived by Grace and two daughters. "We had twenty-five wonderful years together. We'll always have those memories," Grace said.

MAURICE W. SCHUTT

Maurice Schutt was a lifelong resident of Nashville. He was appointed to the Nashville Fire Department as a sub-fireman in February 1930 and made a regular fireman the following May. He was promoted to lieutenant in 1936 and captain in 1939. In 1944 he became drill master.

Many of the Fire Department's new training services were the product of his work.

Mr. Schutt suffered a heart attack at the department training center on June 13. 1963.

VERNON KEEL

Chief Vernon Keel was a veteran of the fire department for 32 years. He suffered a heart attack that claimed his life on March 5, 1971.

THURMOND GENTRY

Captain Thurmond Gentry suffered a fatal heart attack while cleaning fire apparatus at the fire hall on May 24, 1979.

EDDIE MONTGOMERY

The Nashville Fire Department was stunned by the death of Eddie Montgomery. Fire department records show that a call was answered to 140 North First Street on January 14, 1980, that claimed his life.

Three construction workers were overcome by carbon monoxide gas in a sewer line 800 feet from a six-foot opening at ground level. Two construction workers who escaped from the sewer line said that three fellow workers were trapped by fumes when the welding equipment they were using apparently consumed the available oxygen. The two men emerged by way of the six-foot opening and ran to summon the fire department and ambulances.

Eddie Montgomery on Medic 5, equipped with oxygen tank, entered the six-foot opening but was never heard from again until he was brought out unconscious. He was pronounced dead at 11:30 p.m. at General Hospital.

JOHN HALL

John Hall was killed while answering a fire call on October 12, 1979. He worked at Fire Hall 31.

CHARLES LOVELL

Charles Lovell suffered a fatal heart attack on September 19, 1980, while fighting a fire at 12th and Jo Johnson.

NAMELESS HEROES

In the early days of firefighting, a fireman was often judged by how long he could stay in a smoke-filled building with no protective breathing apparatus. Those who could stay for long periods of time were referred to as "smoke eaters".

Many of these men paid dearly for their reputation with heart and lung problems. A bill passed in the Congress of the United States and adopted by the legislature of the State of Tennessee in 1963, placed deaths of firemen caused by heart and lung disease in the "line of duty" category. All firemen today are required to wear modern protective breathing apparatus and protective clothing in all emergency situations.

Records were not kept of those "smoke eaters" who lost their lives due to heart and lung problems. However, we take this opportunity to honor their memory for the sacrifice made by them.